T0276485

System of Systems Engineering

System of Systems Engineering

Edited by **Chester Mann**

New York

Published by NY Research Press,
23 West, 55th Street, Suite 816,
New York, NY 10019, USA
www.nyresearchpress.com

System of Systems Engineering
Edited by Chester Mann

International Standard Book Number: 978-1-63238-432-4 (Hardback)

Printed in the United States of America.

Contents

Preface

The world is advancing at a fast pace like never before. Therefore, the need is to keep up with the latest developments. This book was an idea that came to fruition when the specialists in the area realized the need to coordinate together and document essential themes in the subject. That's when I was requested to be the editor. Editing this book has been an honour as it brings together diverse authors researching on different streams of the field. The book collates essential materials contributed by veterans in the area which can be utilized by students and researchers alike.

Various aspects related to engineering of system of systems have been highlighted in this comprehensive book. It presents and promotes analysis on present applications in the field of system of systems, emphasizing on the significance of the fact that new advancements and area of non-technical as well as technical applications are joining. The aim of this book is to develop an efficient platform for interaction among different types of theory developers and practitioners involved in utilizing the systems engineering approaches and system thinking at the scale of escalated complications and advancing computational answers to such systems.

Each chapter is a sole-standing publication that reflects each author´s interpretation. Thus, the book displays a multi-facetted picture of our current understanding of application, resources and aspects of the field. I would like to thank the contributors of this book and my family for their endless support.

Editor

1

From System-of-Systems to Meta-Systems: Ambiguities and Challenges

G. Reza Djavanshir[1], Ali Alavizadeh[1] and M.J. Tarokh[2]
[1]*Johns Hopkins University,*
[2]*K.N. Toosi University of Technology,*
[1]*USA*
[2]*Iran*

1. Introduction

While the term system-of-systems (SOS) is widespread and generally recognized within academia and industry, there is still confusion about its definition. For example, there are many different interpretations of system-of-systems (Sauser, and et at., 2009). A sample of different definitions is provided by Jamishidi (2005). So far, little research has been done to provide standard definitions of its characteristics. Many prominent researchers including Dagli, Kilicay-Ergin, (2009), Buede (2009), Keating (2009), Eisner (1997), and Djavanshir (2005) accept that systems-of-systems are meta-systems that exhibit meta-systemic behaviors. The meta-system provides the structural mechanism that integrates a system-of-systems as a whole ((Beer, 1979,1981) and Keating (2009)) and prevents it from falling into chaos. This is accomplished by a meta-system's governance system. Therefore, a meta-system is a system-of-systems that has an additional characteristic, called a governance system, which integrates the system-of-systems' components, and provides balanced operations among them in order to achieve a common mission and strategy. According to our studies and understanding (Beer 1979, 1981, Esiner (1997), Keating (2009), Dagli and Kilicay-Ergin (2009), Buede (2009), and Djavanshir, et. al., 2009), if any system-of-systems possesses two properties, namely, (1) evolutionary process between the system-of-systems and its component enterprise systems and (2) passion of integrated centralized governance system, the system-of-systems is called a meta-system. In this chapter we will call a system-of-systems with these two characteristics a meta-system. Furthermore, the concept of meta-system provides powerful means to a better understanding of the so-called system-of-systems' nature, characteristics (Esiner, 1997, Djavanshir, and et. al., 2009, Klir, 1985, Kawakek, 2002, and Buede, 2009), behaviors (Keating, 2009), and finally, its structure (Dagli, Kilicay, Ergin, 2009). Furthermore, accepting the fact that meta-systems are the extended version of meta-systems with these properties, enables researchers to apply well-researched, standardized and accepted definitions, behaviors, and characteristics of meta-systems in studying, understanding, designing, and deploying the so-called systems-of-systems.

Therefore, a meta-system is an extended and robust version of a system-of-systems. In this book chapter we attempt to define it.

A meta-system is a complex system-of-systems with a centralized governance structure that coordinates the operational behaviors of the component systems and provides the strategic framework that guides the component systems to the achievement of their shared emergent mission. The component systems of meta-systems are composed of technological artifacts and informational, organizational, managerial, and human elements; these heterogeneous elements are integrated together to create emergent capabilities and capacities for achieving their shared function(s). Meta-systems possess a governance system that controls, which are described as follows:

1. **Infrastructure system** can include electric power grids, roads, airport facilities, supply chains, tools, assembly lines, technological artifacts, and all other resources.
2. **Communications system** contains various multimedia networks such as, voice, data, and video communication networks; the internet and intranet; and learning channels that provide knowledge gain and accumulations that are essential to both meta-systems' operations and its effective functioning within uncertain and changing environments. The communications system also provides error detections, feedback, fault isolations, and correction mechanisms by continuously reexamining the adequacy of its design (Keating, 2009). A communication exchange between a meta-system's components not only make achieving the function possible, but they also enable the components to evolve and adapt to each other. Communications system also help the emergence of a self-organizing structure in chaotic situations (Keating, 2009).
3. **Governance system** is composed of people, processes, organizations, and products that provide interface protocols for those involved in the design and operation of meta-systems. The governance system enables the meta-system to provide smooth operations of components. Also, it control the the overall operation of achieving the meta-systems' function. However, it does not manage the daily operations of the component systems. That is, flexibilities are provided to all components in choosing the tactical executions of their functions. A meta-system's ultimate role is to provide a seamless design of its three main system-components, in order to create a shared capability to achieve the meta-system's function.

   Heterogeneous elements such as, people, institutions, organizations, information, and technologies create various systems that make up the components of a meta-system. In other words, the components of a meta-system are systems whose elements come from three integrated networks. Therefore, component systems are composed of heterogeneous elements of technologies, tools, processes, people, organizations, information and communications networks, and resources.

This chapter will be composed of five sections: Section 1 will provide an introduction, Section 2 will discuss the characteristics of meta-systems, and Section 3 will provide the description of, and the rational for, a centralized governance mechanism. The solicited opinions of a survey of experts and practitioners' views on systems-of-systems and meta-systems will be described in Section 4. Finally, Section 5 will provide a conclusion and recommendations for future research.

2. Characteristics of meta-systems

Building on the definitions provided by Sage, Cuppon; Keating(2009), Dagli and Kilicay-Ergin (2009), Eisner (1997), and (Djavanshir, et al., 2007), we came up with the following critical characteristics:

2.1 Different elements

Meta-systems are composed of systems whose elements can be: technological artifacts, information, and communication channels; energy generators and transmission systems; or organizations, people, and processes.

2.2 Constrained Autarky of distributed enterprise systems

According to Webster's dictionary, Autarky means self-sufficiency and managerial independence. Autarky of distributed enterprise (component) systems in the context of meta-systems means that the component systems are separate and autonomous systems, yet they are integrated and combined together by the meta-systems, and they exist for the purpose of serving the shared mission of the meta-system (Luhmann, 2003 and Kauffman, 1994). While they have operational and managerial tasks (Wells and Sage, 2009 and Eisner and Marciniak, 1991), there are still resources, financial, and capital flows between the distributed enterprise systems and the governance system of the meta-systems. Constrained autarky of distributed systems implies that these systems do not possess full, independence from the meta-systems that govern them by way of the design of these systems, which is to ensure the effective accomplishment of their shared function and to prevent the operations of the meta-systems from falling into chaos. Moreover, constrained autarky means that the variable autonomies of distributed systems range from full autonomy to full dependency. In other words, the autonomy of each distributed system is a function of variables such as, skills and talents the criticality distributed systems' operations to the overall mission, operation of a meta-system, and the amount of investments spent on the systems. For example, distributed systems do not have decisions that are critical to the survival or failure of the meta-systems, they cannot make decisions, nor can a distributed system decide to ignore the decisions that are accepted by the entire meta-system. However, depending on the mission, distributed systems are either autonomous in making small decisions or they are autonomous in managing their programs and running their daily operations in terms of scheduling, maintenance, and acquisitions.

2.3 Topological dispersion of distributed systems

Component systems can be dispersed in large geographic areas. These systems collaborate with each other and are linked to the governance mechanism through constant communication and the flow of resources, information, and strategic decisions to fulfill the meta-system's overall mission.

2.4 Governance system

Meta-systems possess a governance system that provides an overall control and oversees the system's overall function and operations. The role of a governance system is to control and manage the component systems' behavior.

Governance structure also articulate the functions and overall mission of the meta-systems. Additionally, it facilitates the teamwork and effcient processes throughout the entire system. In order to respond to the changing environment and to effectively achieve its goals and missions, a meta-system must have a control system that manages the changes in its overall functional behavior. Strategic control is not the detailed change strategy, but rather it articulates the meta-system's mission and functions and provides a general description of how the system can accomplish its stated function. Strategic control is based on the assumption that the detailed knowledge about a change or design strategy is part of the change or design process. Therefore, it provides a flexible framework for implementation tactics.

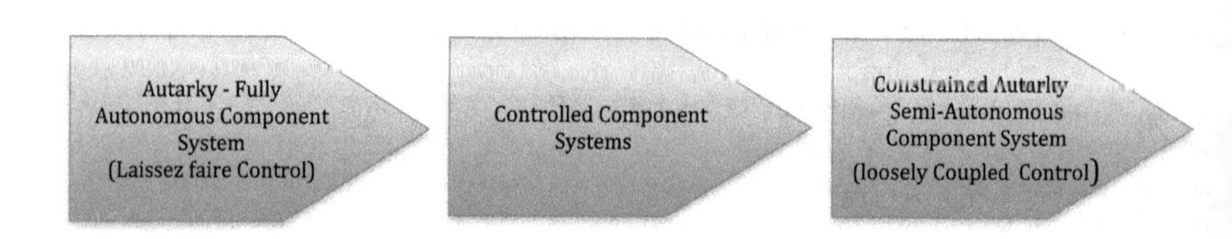

Fig. 1. Continuum of Constrained Autarky (Variant-Autonomy) of Component Systems

2.5 Evolutionary process

The environment in which meta-systems are located is always uncertain and evolving, therefore, the requirements specification of meta-systems is always evolving and changing. The component systems are created for and integrated by means of the meta-system, while the meta-system itself exists and is created by means of its interacting component systems. The process of the creation of a meta-system and its component systems, results in self-organization of the entire meta-system.

Meta-systems are developed through an evolutionary process, where the components of the meta-system are modified in response to the changing environment. As the environment changes, meta--systems' and configuration change as well. Therefore, a meta-system's, configuration, and final design are always evolving (Well and Sage 2009). In other words, as the environment evolves, the meta-system redesigns itself to adapt and respond to new conditions in a evolutionary manner. Evolution means the environment changes, the meta-system's entities adapt to their environment over time.

2.6 Emergence property

There are two related forms of emergence or type II properties (Gharajedaghi, 1999) in meta-systems. The first type is intentional and by design, where new capabilities, behaviors, and properties emerge from the process of structural interactions among the meta-system's components. The second type is the unintended consequence of not knowing about the emergence in advance (Keating, 2009). This unpredictable behavior comes from the process of the interactions of the meta-system within an uncertain and unpredictable environment.

With respect to the changing environment and the emergence of unintended consequences, the detailed design and specifications of a meta-system should not be specified in advance of its operation. Rather, the knowledge and information about the meta-system's design should be part of its design, operations, and implementation process. In other words, the system design should be based on the law of *minimum specifications* (Keating, 2009). Otherwise, the detailed specifications and design increase the system's complexity, which creates a structural sclerosis that restricts the meta-system's agility and its responsiveness to the evolving uncertain environment and emergent unintended consequences. Furthermore, it restricts the meta-system's capacity to self-organize and self-produce based on the contextual information. Therefore, meta-systems should be designed with minimum specifications.

3. The governance system

Governance controls the collective actions overall function of the component systems in the meta-system. As shown in Figure 2, the governance structure is integrated but centralized entity in direct contact with all component systems of the meta-system that exchanges information. In meta-systems, centralized governance system mechanism is needed. It does not manage component systems in terms of providing control. However, it formulates the overall design of the meta-system. It provides control, and coordination of efforts to achieve the mission. and to prevent the system from falling into chaos.

Meta-systems are complex constructs. Their complexities, coupled with the uncertainty of the environment. DeRosa (2001) emphasizes the need for a governance mechanism in large-scale enterprise systems that are composed of human, informational, organizational, and technological elements. The governance mechanism is used to:

1. Define the goal that should be achieved through the meta-system, as well as the capabilities expected from component systems, in order to achieve the desired outcome. Component systems should collectively provide the expected functions to succeed in achieving their mission.
2. Facilitate the flow of information, to facilitate operations
3. Facilitate the creation of a environment that maintains the control of meta-systems during internal and external changes and turbulence (Keating, 2009). Balance and the maintenance of balance, is achieved through adjustments to changes, shifts, and disturbances (Keating, 2009).
4. Formulate the meta-systems' governance functions to avoid chaos. In changing environments, meta-systems' design are always evolving, and their development and deployments evolve with the changes in the environment and in requirements. Therefore, meta-systems' and integration proceed according to uniform standardized design integrations model. The design models and the overall fitness rules are shaped by the efforts of the meta-systems' governance system and its component systems. The governance system creates rules and structures which will in turn aid in the achievement of the meta-system's functions. Even if the individual component enterprises and their managers pursue their own self-interests, the Governance system should create an environment for the functioning of all members of the meta-system (McCarter and White, 2009).

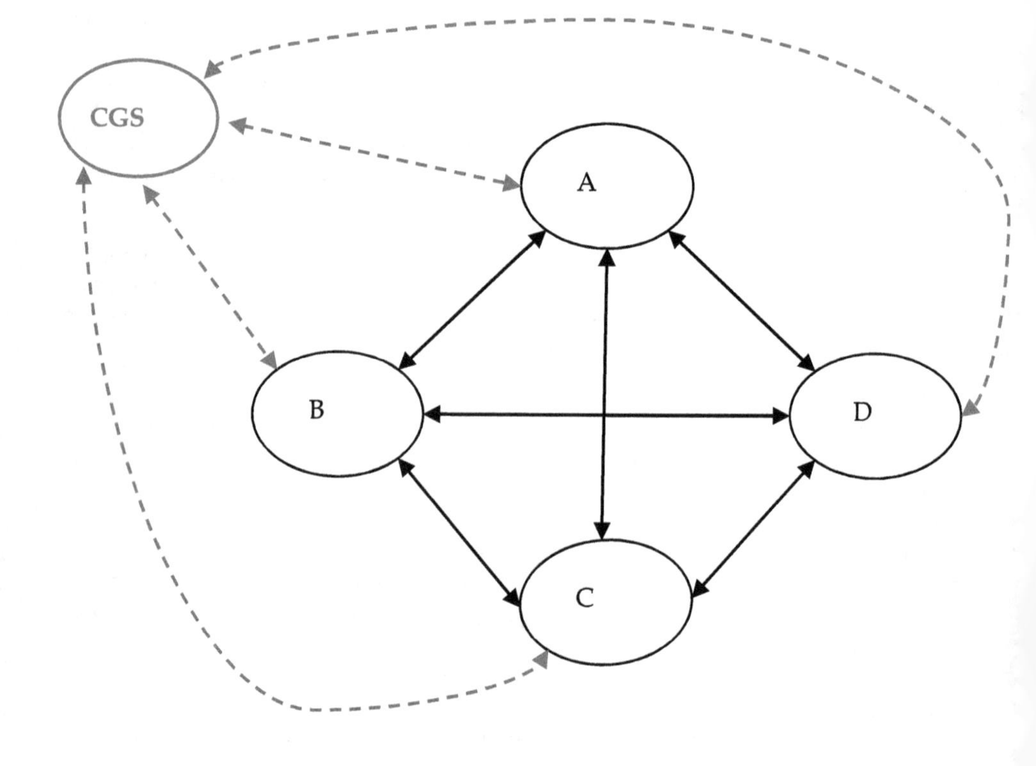

Fig. 2. Illustration of a Governance Structure

4. Experts' views on meta-systems

To provide a comprehensive view of system-of-systems, we also conducted literature research with experts in order to solicit experts' opinions of system-of-systems and meta-systems. A more in-depth reasoning behind the definitions was needed to establish a true understanding of what a system-of-systems is, the characteristics that define it, as well as the differentiation between a system-of-systems and a meta-system. This section explores the results of selected surveys. It is important to note that, given the short timeframe within which this research was conducted, only literature search was used. Therefore, the obtained results are certainly not enough to provide a "representative sample." The majority of literature articles are, however, were written by subject matter experts in the field. The research results were gathered from average system-of-systems experts with special practical knowledge of systems-of-systems; in other words, not academic subject matter experts, but also those with professional perspectives from the field. The information that was obtained is described in this section, but the authors welcome additional researchers to continue this effort. The results obtained to date are quite intriguing, and rather contradictory.

To provide a focused and coherent structure to our interviews, two questions were asked and searched in some instances where time or interest did not permit the respondent to complete the entire questionnaire. It was necessary to shorten the scope of the questions for some respondents and us in order to obtain timely responses. In one instance, the response was limited to an even smaller subset. The questions were as follows:

1. What is your definition of a system-of-systems?
2. What is your definition of a meta-system?

Some definitions (Djavanshir, et, al., 2007) include central control of the overall system and use the term meta-system to describe a system-of-systems while others do not recognize the concept of centralized control (Maier, 1998). Subject matter experts in the field express different opinions on SoS control ranging the gamut, including control as a necessity (Djavanshir, et, al., 2007), control as an impossibility (Maier, 1998) and control as a possibility depending on SoS hierarchy. We believe that the concept of centralized control does hold merit as it pertains to the systems environment because a centralized control structure is required to maintain order of the system. Whether the term meta-system is synonymous with SoS, however, is a subject requiring additional research.

Experts believe that evolution of systems has led to the development of the system-of-systems concept (Djavanshir, et. al., 2007) System-of-systems generally include various heterogeneous systems such as existing systems and new systems in some instances, to provide a particular functions or service. Thus, the challenges and complexities involved in designing such systems have given birth to a new field of systems engineering known as SoSE or System-of-Systems Engineering. Researchers are continuously researching the methodologies and processes currently in practice for SoSE. Two main groups are leading the charge: National Center for System-of-Systems Engineering (NCSOSE) and the System-of-Systems Engineering Center of Excellence (SoSECE). These groups are at the forefront of SoSE and have taken leadership roles in standardizing the discipline. Nevertheless, this new field has brought on some challenges, especially regarding its design. Due to the complex nature of SoS, emergent factors, and the fact that the newly designed system may be a mixture of different existing and new systems that are geographically located, the SoSE design team must be diligent and consider all factors that may affect overall performance of the system.

One thing is clear: the future opportunities of system-of-systems applications such as, in artificial intelligence, are limitless, and are bound to change the world. Thus, SoS applications will become more common, requiring greater understanding and standardization of SoSE design principles to address the demands of the emergent properties of a SoS.

The future of system-of-systems appears endless in this ever-growing age of technology. This concept of large-scale integration can be used to satisfy numerous goals for future projects. This new area will probably be beneficial across many nations, governments, and multinational corporations. It is becoming more widely accepted that system-of-systems will present a real opportunity to future executives of government agencies and industrial companies. New computing paradigms such as Artificial Intelligence (AI) will continue incorporating fuzzy logic into systems-of-systems and will push into mainstream efforts.

There are numerous requirements that could necessitate the advancement and desire for future applications of meta-systems. Since more companies are multinational, the sharing of information between independent systems could be the key to success for the company as a whole. For example, multinational financial institutions such as banks and investment firms usually operate independently, but the financial markets globally are so closely linked, they must continually come up with ways to better integrate their systems. The top level leadership must ensure they have a control system which allows information to flow smoothly both ways.

The transportation industry, for example, could significantly benefit from a new and improved new meta-system in the future. All independent transportation systems, including air travel, maritime, trains, and buses, could be individual complex components of a larger Department of Transportation meta-systems. The recent trends towards modular systems development would greatly enhance this capability. The information from this system would be critical in natural disasters and national emergencies. The benefits are also weighed down by drawbacks. Even though these systems are independent from each other, any negative effect on one could dramatically affect the others. If there were, say, grounded aircraft, this could increase the traffic for trains and buses dramatically.

The opportunities seem endless but the near term focus would seem to be in the Department of Defense (DOD), large government agencies and multination corporations. Projects could include such things as future combat systems for the DOD, improvements to the FAA's aviation program, NASA's space and satellite program, Global weather forecasting, and sharing of information and resources between independent branches of large corporations. This could also prove beneficial to large-scale projects for the federal governments.

The experts and practitioners' responses to these questions were as follows:

Andrew Sage: His definition of a system-of-systems, and reiterated in his communications with one of the authors, does not include any mention of control of the overall system. He also uses the term "Systems Family." He does not, however, use the term meta-system. His understanding of a meta-system as it is defined by Renee Stevens at MITRE, is that it is essentially the same as a system-of-system (SOS). Sage believes that there needs to be a way to govern and manage a SOS, but that diverges into the arena of Federation of Systems efforts. He has not considered the inclusion of a centralized transition control strategy in a SOS.

Maier: He does not recognize the term meta-system, since it is not standardized. His definition conflicts with the concept of control needs in a system-of-systems, stating, "To me, SOS and lack of central control are synonymous. I don't know of any single best practice for stability." In general, "stability" is not a well-formed concept for complex things. There is technical stability, which is how a control theorist would describe stability. He delves deeper into these issues in his papers, "Architecting Principles for Systems-of-Systems" and "On Architecting and Intelligent Transport Systems." This view is quite interesting.

Jamshidi: He defines a system-of-systems as, "a super-system consisting of an integration of an emerging set of heterogeneous systems required to work together for a common purpose, i.e. increased robustness, performance, cost, etc." He believes that meta-systems are more general than systems-of-systems, though some meta-systems may also be systems-of-systems, as stated below. Meta-systems have several definitions that link the concepts of systems-of-systems with those of meta-systems. A vague indication, suggested by A.M.Gadomski is, "one may assume that meta-systems are ***systems*** composed of the common *properties* of a large class of systems, but not related to its particular *domain-dependent* properties." According to V. Turchin and C. Joslyn, this "natural" definition is not sufficient for the Theory of Meta-system Transition; it is also not congruent with the definition of system-of-systems in Systems Theory. Regarding control, Jamshidi agrees that centralized control of a SOS may be possible, depending on its architecture. He believes a hierarchical architecture would lend itself to such a paradigm. Jamshidi also agrees that the requirements management of a SOS is daunting, "Modeling of SOS is a very challenging task, if not impossible. However, it is possible to utilize a peer-to-peer approach for data exchange between systems of a SOS, using tools like XML language and discrete-event simulation to actually simulate a SOS without the benefit of a mathematical model. Currently, we are looking at national Instruments' LabView as an alternative approach to simulating SOS."

Crossley: William Crossley's definition is as follows of SOS, "A system-of-systems is any collection of systems, each of which is capable of its operation, that must interact to achieve their purposes or gain value none can fully realize alone. Like a single system, a SOS is a collection of components interacting to fulfill one or more functions. But the constituent systems of a SOS can perform useful functions alone - something components of a single system cannot - and removal of any system from a SOS need not prevent its continued operation."

His thoughts regarding SOS control are that the level of control over a system-of-systems helps in its classification. One of the ideas we have been developing is that the amount of classifications exerted over the component systems is a way to classify a system-of-systems. A system-of-systems, like a battle group or an airline (two examples we have used), could have a fairly "strong" controlling authority. Other systems-of-systems may have very little centralized control; other researchers have attempted to describe the internet/worldwide web as an "uncontrolled" system-of-systems. Mr. Crossley has been researching methods of identifying SOS Engineering problems. He is "looking at how to describe "design" problems in a system-of-systems context using an optimization (or mathematical programming) problem statement. This may only work for a class of SOS problems." This research is very much needed. A scientific or methodical approach to handling SOS Engineering is needed in every aspect, including: the identification of the problem, analysis, requirements management, etc. The tools and methods in place for handling systems engineering are not sufficient for the additional complications of SoSE.

Industry Practitioners: The industry practitioner respondents recognized the terms system-of-systems, meta-systems, and enterprise systems, but were not sure of the differences between them. One suggested that *"system-of-systems is a term that has been recently coined, in the last few years, and refers to net-centric, distributed systems spanning many types of domains."* In general, the terms were not differentiated well. At least one believed that the progress of systems-of-systems would be hindered in the future, *"The drawbacks are that we do not have a good handle on the definition and implementation of System-of-Systems today. The fact that we are dealing with ever-increasing interface standards and poorly engineered COTS solutions threaten to limit their use."*

The surveys confirm the author's beliefs that SOS engineering is in its infancy. Some systems engineers have cursory knowledge of the problem, while others treat SOS problems the same as they treat any other systems problems. It will take time (and standardization) of this industry before the information trickles-down and reaches the field. Standardization will lead to better processes and more efficient methods.

5. Conclusion and recommendations

In this book chapter we provided a definition of a meta-system and its main characteristics. A meta-system provides the structure, processes, and governance mechanism that integrate and synchronize the operational capabilities of SOS. Meta-systems are composed of heterogeneous component systems consisting of: people, technological artifacts, infrastructure, resources, support systems, information, organizations, and regulative, normative, and cultural cognitive institutions. Meta-systems have uncertain environment changes (sometimes with high velocity), incomplete and variable specifications, and an elastic boundary.

There are also symbiotic and commensalist relationships between a meta-system and its component systems. Symbiotic and commensalist relationships mean that intertwined, interdependent or partially interdependent entities *help each other* (symbiotic relationship) and *use each other* (commensalist relationship) respectively (Eisenhardt and Galunic, 2000). Furthermore, meta-systems and their component systems not only co-create each other, but they also co-adapt, collaborate, and co-evolve.

We also emphasized on the importance of a loosely centralized governance mechanism that governs (not manages or rigidly controls) the overall operation of meta-systems and prevents meta-systems from falling into chaos. It also balances the opposing tendencies within meta-systems. Governance mechanisms provide balance among the opposing tendencies, guidance and policies that facilitate cooperative behaviors, and guidance and policies that further the emergence of self-organizing behavior out of complex and chaotic situations.

In this book chapter, we also provided a survey of experts and practitioners' views on system-of-systems, meta-systems and their differences.

For future research, it is recommended that the concepts of governance mechanism and various degrees of the autonomies of its component systems be examined and their governances also be studied.

Additionally, the concept of the complex system (self-organizing, where a whole exists for and is created by its parts and vice-versa) can further examine meta-systems.

6. Acknowledgement

The authors would like to thank Ms. Katherine L. Hudak of Johns Hopkins' Carey Business School for her insightful editorial work.

7. References

Beer, S. (1997). The Heart of Enterprise, John Wiley

Buede, D., M. (2009). The Engineering Design of Systems, 2nd Ed. John Wiley. Hoboken, N.J.

Dagli, D., H., and N., Kilicay-Ergin. (2009). "System of Systems Architecting." In System of Systems Engineering, Innovation for the 21st Century. Jamshidi, M., ed. Hoboken, N.J. John Wiley, pp. 77-100.

DeRosa, J., K. (2011). "Introduction." In Enterprise Systems Engineering. Rebovich, G., Jr., and B.E. Whites, eds. Boca Raton, FL: CRC Press, pp. 1-30

Djavanshir, G.R., Khorramshahgol, R. and J. Novitzki. (2009). Critical Characteristics of Meta-systems: Toward Defining Metasystems Governance Characteristics. IEEE, ITPro. May/June Issue, pp. 31-34.

Eisner, H. (1997). Essentials of Project and Systems Engineering Management. NJ: John Wiley, p. 214.

Eisenhardt, K., and D. C., Galunic. (2000). Coevolving At Last a Way to Make Synergies Work. Harvard Business Review. January-February, pp. 83-101.

S. Kaufffman. (1995). At Home in The Universe. NYC. NY: Oxford University Press, pp. 274-275.

Kawakek, P., and D.G., Wastell. (2002). "A Case Study Evaluation of the Use of the Viable System Model in Information Systems Development." In Information Systems Evaluation Management. Van Grembergen, W., ed. London, UK. IRM Press, pp. 17-34.

Keating, C. (2009). "Emergence in System of Systems." In System of Systems Engineering, Innovation for the 21st Century. Jamshidi,M., ed. Hoboken, N.J. John Wiley, pp. 169-217.

Keating, C. et al., (2003). "System of Systems Engineering," Eng. Management J., vol. 15, no. 3, pp. 36–45.

Klir, G. (1985). Architecture of Systems Problem Solving, PlenumPress, p. 305.

Maier W.M. (1998). "Architecting Principles for System-of-Systems," Systems Eng., vol. 1, no. 4, pp. 267–284.

McCarter, B. G., and B. E. White. (2009). "Emergence of SOS, sociocognitive aspects." In System of Systems Engineering, Principles and Applications. Jamshidi, M., ed. Boca Raton, FL: CRC Press, pp. 71-105.

Sage, A.P., and C.D. Cuppan. (2001). "On the Systems Engi- neering and Management of Systems and Federation of Systems," Information, Knowledge, and Systems Management, vol. 2, no. 4, pp. 325–345.
Sage, A.P. (1992).Systems Engineering, N.J: John Wiley.
Sage, A. P. (2003). "Conflict and Risk Management in Complex System of Systems Issues," IEEE Int'l Conf. Systems, Man, and Cybernetics, IEEE CS Press, 5. pp. 3296–3301.
Stevens, R. (2011). Engineering Mega-systems. CRC Press. Boca Raton, FL.

2

System of System Failure: Meta Methodology to Prevent System Failures

Takafumi Nakamura[1] and Kyoichi Kijima[2]
[1]*Fujitsu Fsas Inc.,*
[2]*Tokyo Institute of Technology,*
Japan

1. Introduction

The purpose of this chapter is to propose a meta methodology to promote engineering safety by learning from previous system failures. The predominant worldview in IT engineering is that systems failures can be prevented at the design phase. This worldview is obvious if we examine mainstream, current methodologies for managing system failures. These methodologies use a reductionist approach and are based on a static model (Nakamura & Kijima, 2007, 2008a). It is often pointed out that most such methodologies have difficulty coping with emergent properties in a proactive manner and preventing the introduction of various side effects from quick (i.e., temporary) fixes, which leads to repeating failures of similar type. There are many examples of similar system failures repeating and of negative side effects created by quick fixes. Introducing safety redundant mechanisms does little to reduce human errors. As pointed out by Perrow (1999, p. 260), the more redundancy is used to promote safety, the greater the chance of spurious actuation; "redundancy is not always the correct design option to use." While instrumentation is being improved to enable operators to run their operations more efficiently and certainly with greater ease, the risk would seem to remain about the same. The main reason for this situation is that current methodologies tend to identify a system failure as a single, static event, so organizational learning tends to be limited to a single loop rather than a double loop in rectifying the model of the model (i.e., the meta model) of action (i.e., the operating norm). This indicates that we need a meta methodology that can manage the dynamic aspects of system failure, by ensuring the efficacy of its countermeasures through the promotion of double loop learning.

In this chapter, we propose a meta methodology called System of System Failures (SOSF), along with a system diagnostic failure flow, in order to overcome the current methodologies' shortcomings. We also demonstrate this meta methodology's efficacy through an application in IT engineering.

In the next section, we explain the current troubleshooting techniques' features and limitations with respect to certain aspects of system failures. Section 3 describes the three key features required in order to overcome these limitations, as well as SOSF, which actually overcomes the limitations. In section 4, we propose the actual application scenario that fully utilizes SOSF to promote double loop learning, or total system intervention for system failure (TSI for SF). The SOSF and related methodologies are used in the course of the subsequent discussion and debate to agree upon who is responsible for the failure and to identify the preventative measures to be applied. In section 5, an application example in information and communication technologies engineering demonstrates that using the proposed "TSI for SF" helps prevent future system failures by learning from previous system failures, followed by a concluding discussion of a efficacy of the SOSF and three actions were identified for preventing further system failures: close the gap between the stakeholders, introduce absolute goals and enlarge system boundary.

2. Limitations of current troubleshooting techniques

The predominant technology of current ICT troubleshooting is based on a predefined goal-seeking model. van Gigch (1991) points out the main shortcomings of system improvement in this model, as follows. (1) Engineers look for causes of malfunctions within the system boundary. The rationale of system improvement tends to justify systems as ends in themselves, without considering that a system exists only to satisfy the requirements of larger systems in which it is included. (2) Engineers seek to restore systems back to normal. A lasting solution cannot result from an improvement in the operation of a present system. An improvement in operations is not a lasting improvement. (3) Engineers tend to hold incorrect, obsolete assumptions and goals. It is not difficult to find organizations in which the formulation of assumptions and goals has not been explicit. Fostering system improvement in this context is senseless. (4) Engineers act as "planner followers" rather than as "planner leaders." Another manifestation of the problem of holding incorrect assumptions and pursuing the wrong goals can be traced to different concepts of planning and of the planner's role. In the context of system design, the planner must be a planner leader, planning to influence trends, instead of a planner follower, planning to satisfy trends.

This chapter focuses on system failure aspects that current methodologies cannot manage properly in the sense pointed out by van Gigch. To summarize, these aspects are soft, systemic, emergent, and dynamic; i.e., they accommodate multiple stakeholders' worldviews (Checkland, 1981; Checkland & Holwell, 1997).

Technology is changing faster than engineering technology can treat system failures. The growing increase in CPU power versus price is well known in the form of Moore's law. Moreover, the numbers of stakeholders in computer systems is getting bigger and bigger. For computer architects, the stakeholders should encompass clients of clients (i.e., end users) in order to satisfy ICT system owner's requirements. ICT system provider should focus on the dynamic aspects of end users and ICT system owners (e.g., through capacity planning of web banking system design), as well as on computer components (HDDs, CPUs, etc.) supplied by various vendors in order to implement synthesized functions. The environmental changes surrounding ICT systems, in terms of speed and complexity, are increasing over time. The problem is that once a system failure happens under these circumstances, it is extremely difficult to identify the real root cause. Most troubleshooting methodologies view system failures as resulting from a sequence of events. Furthermore, they focus mainly on the technical aspects of system failures. These models are only suitable for a relatively simple system with unitary participants from a technical perspective.

The following four key features are commonly pointed out for the current troubleshooting methodologies surrounding ICT system environments. Explanations of system failures in terms of a reductionist approach (i.e., an event chain of actions and errors) are not very useful for designing improved systems (Rasmussen, 1997; Leveson, 2004). In addition, Perrow (1999) argues that the conventional engineering approach to ensure safety - building in more warnings and safeguards – fails because system complexity makes failures inevitable.

1. Current methodologies are technically well established (e.g., ISO and IEC standards) but are not always helpful for understanding the real implications of countermeasures and whether they are real solutions or merely tentative fixes from outside the technical arena. Moreover, most methodologies are based on a reductionist worldview.
2. The current troubleshooting mainstream applies cause-effect analysis (or event chain analysis) to find out real root causes. Forward sequences (as in FMEA or event trees) or backward sequences (as in fault trees) are often employed (IEC 60812 (2006), IEC 61025 (2006)). Toyota has a corporate slogan suggesting to "ask why five times" to reach root causes. This promotes finding "what" in order to seek counter measures to the problem. This approach, however, tends to become a victim-finding tool for blaming a specific person or group rather than finding a real root cause.
3. The enormous speed of technological advance causes various misunderstandings between ICT system stakeholders. This responsibility disjunction cannot be managed properly with current methodologies.
4. Improvement of the deviation from operating norm is bound to fail, as van Gigch (1991) points out that the treatment of system problems by improving the operation of existing systems is bound to fail. Current troubleshooting methodologies focus on the following main problems:

- The system does not meet its established goals.
- The system does not yield predicted results.
- The system does not operate as initially intended.

The basic assumption of improvement is that the goal and operating norm are static and predetermined at the design phase and are based on hard systems thinking.

The above four features hinder examination of system failures from a holistic viewpoint, making it impossible to manage the soft, systemic, emergent, and dynamic aspects of system failures.

3. Double loop learning and System of System Failures (SOSF)

3.1 Double loop learning and three key success factors for new methodology

To overcome the current methodological shortcomings discussed above, we need to promote double loop learning. The most important key success factor is the ability to ask a question with respect to a current operating norm (i.e., a mental model). Skill in double loop learning should enable people to question basic assumptions, which leads to modification of their mental models (Fig. 1) to create action producing desired goals, rather than simply modification of their actions under current mental models (Argyris & Schoen, 1996; Morgan, 1986; Senge, 1990).

Double loop learning should influence all three layers listed in Table 1: reality is for changing actions, model is for changing desired goals, and meta is for modifying mental models. Figure 1 explains single and double loop learning in a multi-stakeholder environment based on a double loop learning model (Morgan, 1986). The dotted line in Fig.

1 indicates one specific stakeholder for achieving a goal. The one stakeholder alone is not enough to overcome current methodological shortcomings. We should thus expand double loop learning to account for a multi-stakeholder situation. Under this situation, there are three key success factors for overcoming current methodological shortcomings. First, there should be common language among the stakeholders' mental models (i.e., the mental model box in each stakeholder's domain in Fig. 1). Otherwise, the failures caused by stakeholders' mental model gaps will not be resolved effectively. Second, there should be a meta methodology (i.e., the meta model box in Fig. 1) to promote double loop learning. This meta methodology should be unique between stakeholders; otherwise, the mutually exclusive and collectively exhaustive (MECE) nature of countermeasures is hard to achieve. Therefore, there is only one meta model box in Fig. 1, and it is shared among stakeholders. Third, there should be failure classes based on the origin of a failure. This is essential to ensure the efficacy of countermeasures. There are three origins of system failures: i) the mental model, ii) a mental model gap between stakeholders, and iii) the meta model. These three origins correspond to failure classes 1 (failure of deviance), 2 (failure of interface), and 3 (failure of foresight), respectively, as indicated in Fig. 1. The following explains the three key success factors in detail.

1. We should have a common language for understanding system failures. It is vital to examine system failures from various perspectives. System safety can be achieved through the actions of various stakeholders. One such common language was developed by van Gigch (1986) for taxonomy of system failures. There are six categories of system failures, namely, failures of i) technology, ii) behavior, iii) structure, iv) regulation, v) rationality, and vi) evolution.
2. We should have a meta methodology to ensure that countermeasures are correct and essential rather than just quick fixes that introduce long-term side effects. To redress system malfunctions or a system failure, it is necessary first to translate specific failure events into a model world in order to appraise the nature of reality holistically, then to discuss the system failure's model in the modeling phase (i.e., metamodeling) in order to investigate why the failure happened, what the countermeasures are, and what should be learned in the organizational process so as to avoid further occurrence of the failure. Kickert (1980) explained an organizational structure model corresponding to the organizational purpose and breaking the organizational structure down into three layers: the aspect system, subsystem, and phase system. These layers relate to "what," "who," and "when," respectively. Beer's VSM model (Beer, 1979; 1980) rectifies the organizational process. Systems 1 to 3 are the operational level, and systems 4 and 5 are the meta level for deciding the operating norm through communication outside the system environment. There are hierarchical similarities between Kickert's and Beer's models, as follows:
   - Systems 1 to 3 correspond to the phase system managing "when." These levels ensure internal harmony and maintain internal homeostasis. Systems 1, 2, and 3 represent when an operation should be done, how it is coordinated, and how to maintain corporate management, respectively.
   - System 4 for strategic corporate management corresponds to the subsystem managing "who." This level integrates internal and external inputs in order to chart enterprise strategies (i.e., external homeostasis) and clarifies who should be responsible for those strategies.

- System 5 for normative corporate management corresponds to the aspect system managing "what." This level formulates long-term policies (i.e., planning and foresight) and decides what should be done.

Kickert's organizational model and Beer's VSM model both decompose organization into three layers: reality (i.e., operation), model (i.e., adaptation), and meta (i.e., evolution). The reality and model layers seek to answer "how," and the meta layer seeks to answer "what." This differentiation is crucial to ensure the efficacy of countermeasures. Table 1 summarizes the relations between the organizational structure (Kickert, 1980) and VSM (Beer, 1979; 1980) models.

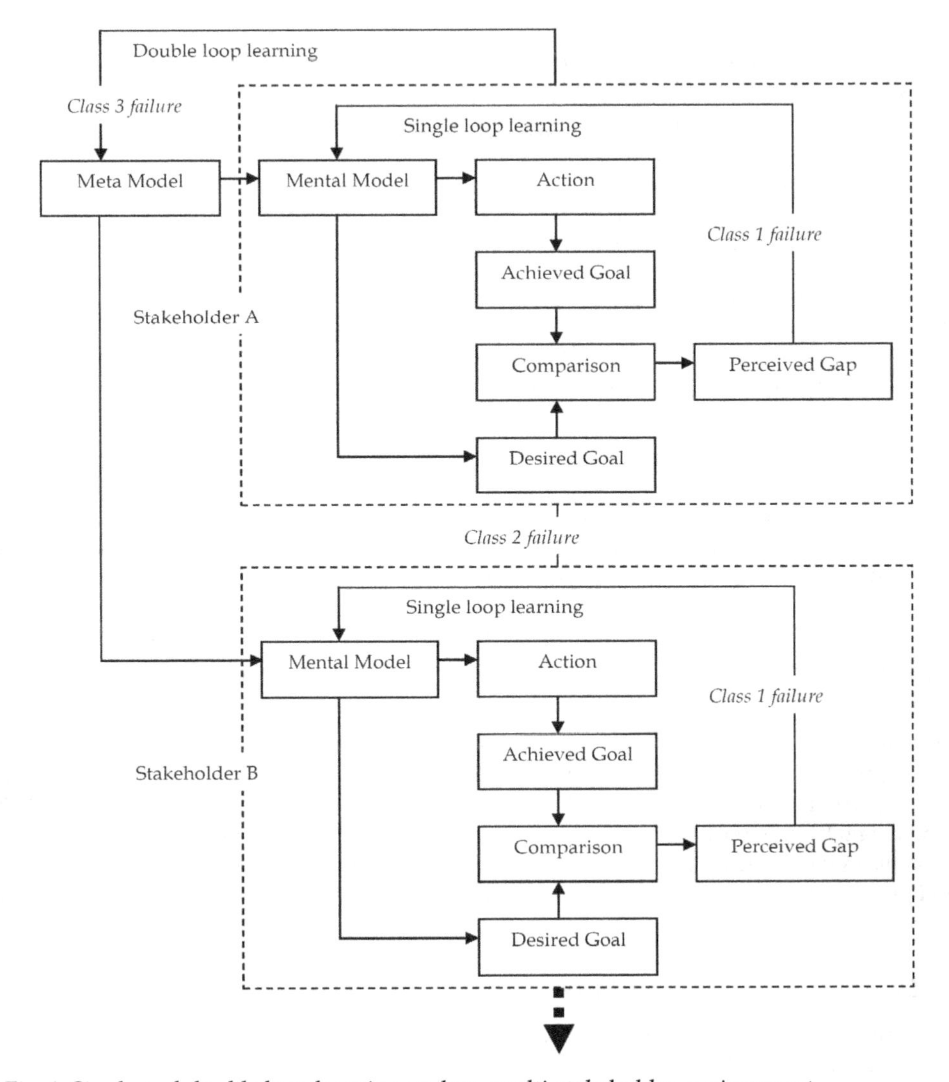

Fig. 1. Single and double loop learning under a multi-stakeholder environment.

	Organization structure	Objective	VSM
Meta	Aspect system: What	Mental model	System 5
Model	Subsystem: Who	Operating norm	System 4
Reality	Phase system: When	Operation	Systems 1-3

Table 1. Relations between the organization structure (Kickert) and VSM (Beer) models.

3. We should be able to specify three failure classes in order to avoid the dynamic aspects of system failures (i.e., erosion of safety goals over time). These failure classes should intentionally be identified in conjunction with the VSM model. They should clarify the system boundary and the nature of a problem (i.e., predictable or unpredictable). The failure classes are logically identified according to the following criteria:
   - Class 1 (failure of deviance): The root causes are within the system boundary, and conventional troubleshooting techniques are applicable and effective.
   - Class 2 (failure of interface): The root causes are outside the system boundary but predictable at the design phase.
   - Class 3 (failure of foresight): The root causes are outside the system boundary and unpredictable at the design phase.

The failure classes thus depend on whether the root causes are inside or outside the system boundary, and a class 3 failure for one person can be a class 1 or 2 failure for other people. Therefore, the definition is relative and recursive, so it is important to identify the problem owner in terms of two aspects: the stakeholder group, and the VSM system (i.e., systems 1 to 5). Unless those two aspects are clarified, failure classes cannot be identified.

It is necessary to recognize the organizational system level in order to rectify the operational norm, because to prevent further occurrence of system failures, it is inadequate to change only systems 1 to 3 (or the phase system for seeking when and how). As pointed out above, current technological models mainly focus on the operational area, and this can lead to side effects resulting from quick fixes. Event chain models developed to explain system failures usually concentrate on the proximate events immediately preceding the failures. The foundation of a system failure, however, is often laid years before the failure occurs. In this situation, the VSM model and Kickert's model serve well for understanding the real root causes.

In a stable environment, control of activities and maintenance of their safety through a prescriptive manual approach deriving rules of conduct from the top down can be effective. In the present dynamic environment, however, this static approach is inadequate, and a fundamentally different view of system modeling is required. Section 3.4 thus describes a dynamic model explaining why fixing failures sometimes introduces unintended side effects and how dynamic understanding contributes to introducing countermeasures that are ultimately more effective.

3.2 System of System Failures (SOSF)

From the above considerations, we now propose a new methodology, called System of System Failures (SOSF), to promote double loop learning and satisfy the above three key success factors. Double loop learning is essential for determining whether operating norms (i.e., mental models) are appropriate (Argyris & Schoen, 1996; Morgan, 1986; Senge, 1990). It also provides a meta methodology for changing mental models so as to overcome system improvement shortcomings (Leveson, 2004; Perrow, 1999; Rasmussen, 1997; van Gigch, 1991), as explained in section 2. Among the meta methodologies proposed in a general context, the System of System Methodologies (SOSM) developed by Jackson (Jackson, 2003) is a typical, excellent example. SOSM's main features are the following: i) a meta systemic approach ; i.e. soft system thinking to foster double loop learning (Checkland, 1981; Checkland & Holwell, 1997), and ii) complementarism by encompassing multiple paradigms (contingent approach by combination of various methodologies from various paradigms, depending on problem situations). Figure 2 shows the framework of SOSM. Various classes of systems thinking are located in two-dimensional space, where the two dimensions are participants and systems. The current troubleshooting techniques discussed in section 2 (i.e., FTA, FMEA, IEC) belong to the unitary-simple domain in SOSM.

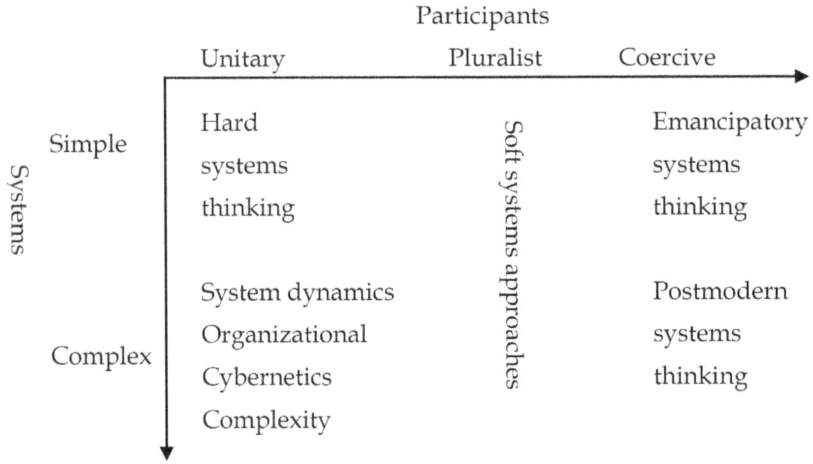

Systems \ Participants	Unitary	Pluralist	Coercive
Simple	Hard systems thinking	Soft systems approaches	Emancipatory systems thinking
Complex	System dynamics Organizational Cybernetics Complexity		Postmodern systems thinking

Fig. 2. Systems approaches related to problem context in the System of System Methodologies (SOSM).

In particular, SOSF is designed by allocating each type of failure from a taxonomy of system failures (van Gigch, 1986) into SOSM space (Fig. 3). There is no coercive domain in SOSF, because the main focus of this chapter is technological systems rather than social systems. The stakeholders for achieving engineering safety are covered fully by the unitary and pluralist domains in SOSM. The allocation of each type of failure from SOSM into SOSF is quite straightforward. The structure connecting SOSM and SOSF is shown in Fig. 4. The left-hand side represents layers of abstraction from reality to methodology to meta methodology. In the realm of system failures, a system failure on the bottom line corresponds to the reality layer. The common language (i.e., the taxonomy of failure) corresponds to the methodology layer. A meta failure (i.e., SOSF) corresponds to the meta methodology layer. Therefore, SOSF is an example of SOSM in the realm of system failure. It is worthwhile to mention the recursive feature of SOSF, depending on the viewpoint of the system. If a target system is broken down into subsystems, each subsystem has its own instance of SOSF. Therefore, a technology failure might be a failure of evolution, one level down, from the viewpoint of the subsystem. Furthermore, this failure of evolution might be a failure of regulation, one level higher, from the viewpoint of the system of systems.

To satisfy the third feature (differentiating the three failure classes) pointed out in section 3.1, we should introduce a third dimension, namely, the failure class. Figure 5 expands two-dimensional SOSF (Fig. 3) into three-dimensional SOSF space, with the addition of the system failure dimension.

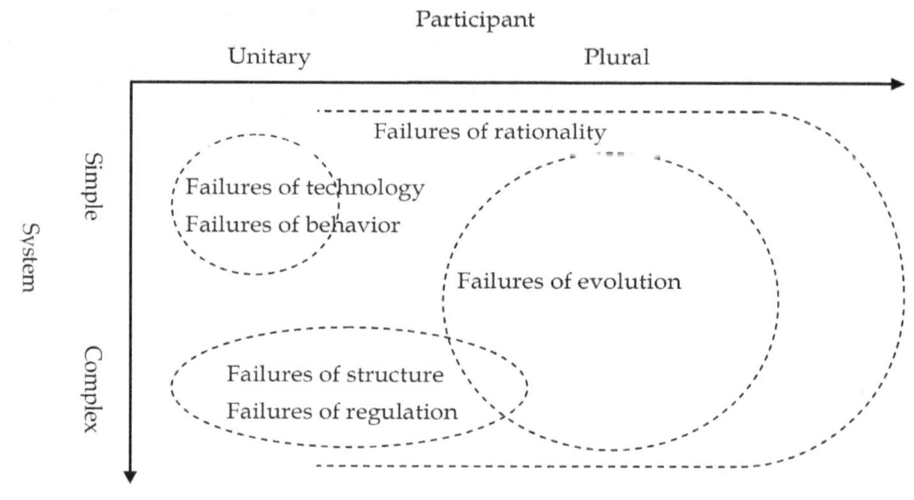

Fig. 3. System of System Failures (SOSF).

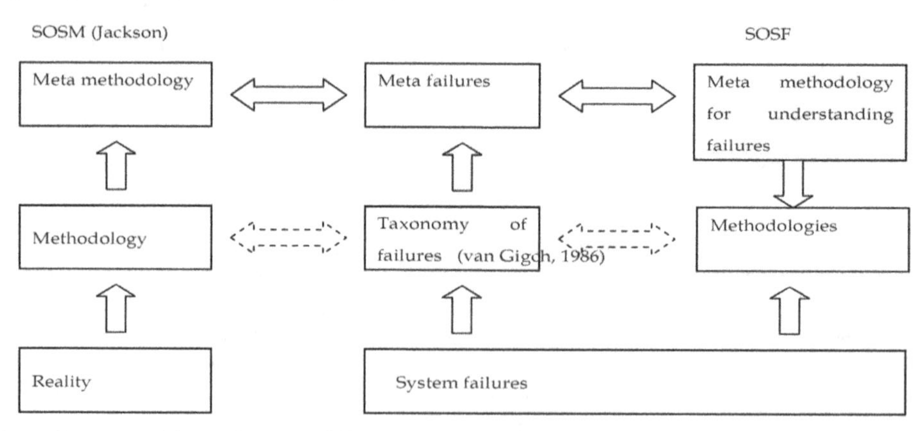

Fig. 4. Meta modeling of system failures and SOSF by using SOSM.

As explained above, because of this recursive nature, it is vital to identify the problem owner in terms of who (i.e., the stakeholder) and where (i.e., the system level in terms of vertical dimension in Table 1).

Table 2 summarizes the general notation of system failures for confirming the mutually exclusive and collectively exhaustive (MECE) nature of the diagnosis, as well as "who," "where," and "what," which stand for the stakeholder, systems 1 to 5, and the failure class, respectively. The horizontal arrows in Table 2 show that at the same system level, stakeholders should be compared in order to identify responsibilities. If a stakeholder is identified, the system level (1 to 5) and objective (what, who, and when) should be identified using the vertical arrows. This ensures the efficacy of double loop learning by changing the model of the model (i.e., the meta model of the operating norm).

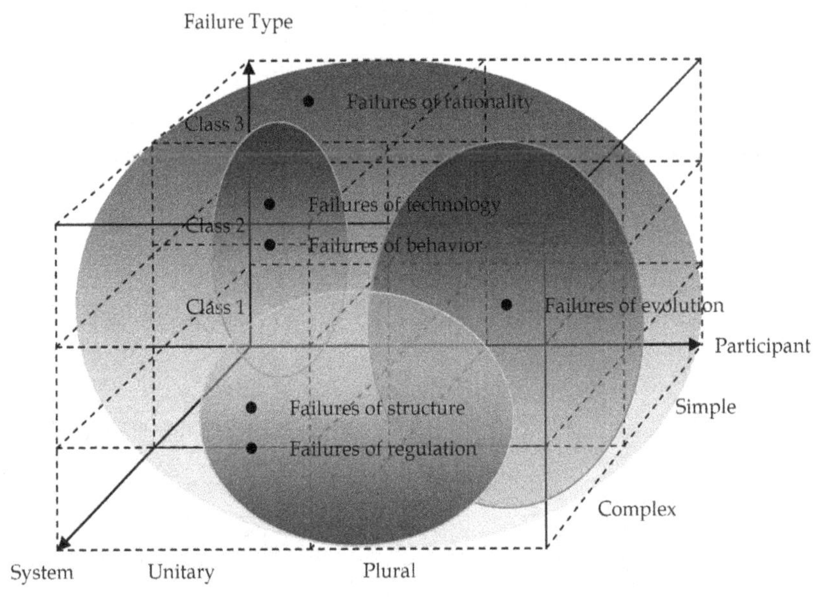

Fig. 5. Three-dimensional SOSF space.

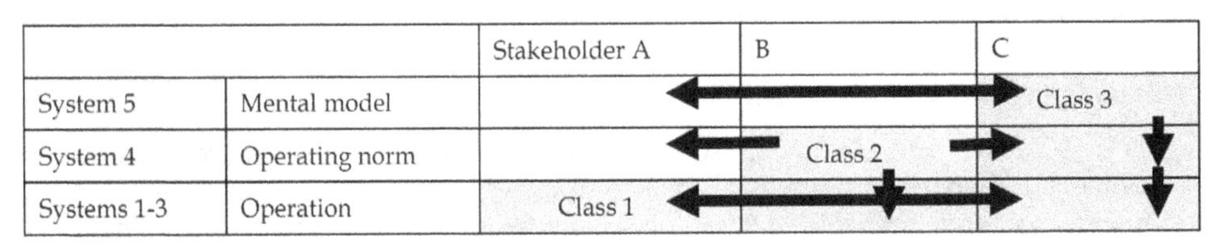

		Stakeholder A	B	C
System 5	Mental model			Class 3
System 4	Operating norm		Class 2	
Systems 1-3	Operation	Class 1		

Table 2. General notation of system failure.

In the next section, we introduce the two new methodologies that cover the SOSF space.

3.3 Failure factor structuring methodology

We propose new failure factor structuring methodology to overcome system failures caused by complex failure factors (Nakamura & Kijima, 2008a). Generally, complex system failures arise from a variety of factors and combinations of those factors. Since these factors often have a qualitative nature, it is important to have a holistic view that reveals the quantitative relationships among qualitative factors in order to construct an effective methodology. The methodology should address complex system failures in terms of obtaining the observations needed to rectify the worldview of maintenance (i.e., double-loop learning). The failure factor structuring methodology (FFSM) should promote double-loop learning through viewing the system in a holistic way. Figure 6 shows a general overview of this methodology, and Table 3 lists the objectives for each phase of FFSM.

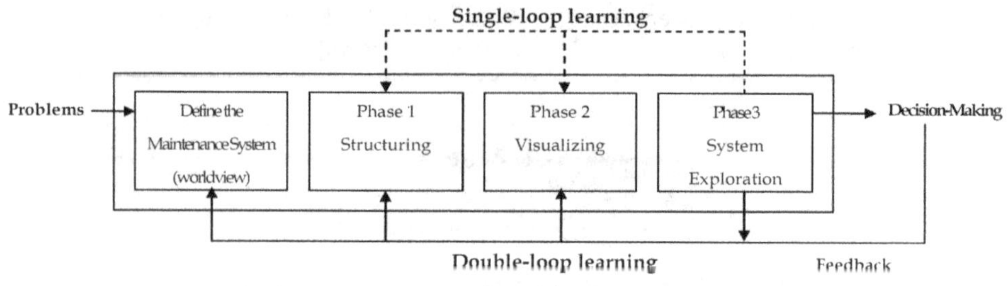

Fig. 6. General overview of FFSM

Phase	Characteristics	Objective
1	• Holistic approach (Structuring factor relationships)	Identify root causes by clarifying relationships among factors
2	• Holistic approach (Grouping factors and problems)	Extract hidden factors underlying complex symptoms by grouping factors and problems
3	• Viewing system from conceptual as well as real-world viewpoint • Double-loop learning	Identify preventative measures for emergent properties by mapping factors into maintenance subsystems

Table 3. Objectives for phases 1, 2, and 3 of FFSM

3.4 System failure dynamic model

We propose new nonlinear systemic model to overcome system failures caused by environmental changes through time (Nakamura & Kijima, 2008b, 2009a). This "system failure dynamic model (SFDM)" is based on system failure class. The frequent occurrence of deviant system failures has become regular but poorly understood. For example, deviant system failure is believed to lead to NASA's *Challenger* and *Columbia* space shuttle disasters (Columbia Accident Investigation Board Report, Chapter 6, pp. 130). This normalized deviance effect is hard to understand from a static failure analysis model. NASA points out the notion of "History as Cause" for repeated disastrous failures (Columbia Accident Investigation Board Report, Chapter 8). These considerations imply usefulness to focus on the dynamic aspects of the cause and effect of system failures rather than the static aspects. Dynamic model analysis is applicable in all technology arenas, including high-risk technology domains like that of NASA. Turner and Pidgeon (1997) found that organizations responsible for a failure had "failure of foresight" in common. The failure or the disaster had a long incubation period characterized by a number of discrepant events signaling potential danger. These events were typically overlooked or misinterpreted and accumulated unnoticed. To clarify that mechanism, Turner and Pidgeon decomposed the system lifecycle from the initial development stage to cultural readjustment through catastrophic disasters into six stages (Turner & Pidgeon, 1977, p. 88). They are Stage I: Initial beliefs and norms, Stage II: Incubation period, Stage III: Precipitating event, Stage IV: Onset, Stage V: Rescue and salvage and Stage VI: Full cultural readjustment. The second stage, or incubation period, is hard to identify due to the various side effects of quick fixes (Turner & Pidgeon, 1997). Therefore the second stage is playing the crucial role to lead catastrophic disaster. System failures have specific features corresponding to these six stages. Class 1 failures occur in the early stages, while Class 2 and 3 failures emerge gradually over time. If we have a way to identify the class of a failure, we can prolong the system life cycle by introducing countermeasures. SFDM should be used periodically to ensure that the system behaves as expected (Reason, 1997, 2003) and that side effects due to quick fixes are prevented.

3.5 Relationships among SOSF and related methodologies

The SOSF meta-methodology overcomes the shortcomings of the current methodologies. The current methodologies (i.e., FTA and FMEA) are reviewed through SOSF and the two new methodologies (i.e., FFSM and SFDM) are proposed to complement the shortcoming of the current methodologies. The relationships among SOSF, FFSM, SFDM, and system failures are illustrated in Figure 7.

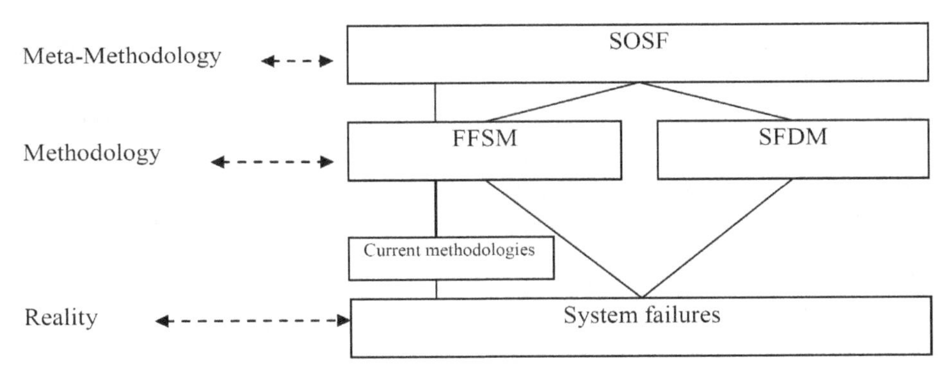

Fig. 7. Relationships among SOSF, FFSM, and SFDM

Table 4 shows the methodology mapping onto SOSF space.

	Within same class	Spread over different classes
Unitary vs. unitary	FTA, FEMA	FFSM
Spread over different domains	SFDM	

Table 4. Methodology mapping to SOSF space

4. Total system intervention for system failure (TSI for SF) methodology as an application procedure

Total system intervention (TSI) is a critical system practice for managing complex and differing viewpoints. In the previous chapter, we introduces meta-methodology called "system of system failures (SOSF)" as a common language among various stakeholders to improve their understanding of system failures. Then we propose the actual application scenario, or "TSI for SF". The SOSF and related methodologies are used in the course of the subsequent discussion and debate to agree upon who is responsible for the failure and to identify the preventative measures to be applied. Flood and Jackson (1991) identified seven principles underpinning the TSI.

First principle: Problem situations are too complicated to understand from one perspective, and the issues they throw up are too complex to tackle with quick fixes.

Second principle: Problem situations, and the concerns, issues, and problems they embody, should therefore be investigated from a variety of perspectives.

Third principle: Once the major issues and problems have been highlighted, a suitable systems methodology or methodologies must be identified to guide intervention.

Fourth principle: The relative strengths and weaknesses of different system methodologies should be appreciated, and this knowledge, together with an understanding of the main issues and concerns, should guide the choice of appropriate methodologies.

Fifth principle: Different perspectives and system methodologies should be used in a complementary way to highlight and address different aspects of organizations and their issues and problems.

Sixth principle: The TSI sets out a systemic cycle of inquiry with interaction back and forth between its three phases.

Seventh principle: Facilitators and participants are engaged at all stages of the TSI process.

Jackson (2006) argues the sixth principle refers to the three phases of the TSI meta-methodology: *creativity, choice, and implementation.* These three phases precede a reflection phase. Therefore, the critical systems practice it embraces is an enhanced version of 'total systems intervention' (Flood & Jackson, 1991), which has four phases: *creativity, choice, implementation,* and *reflection* (Jackson, 2006).

Based upon the seven principles identified by Flood and Jackson (1991), we introduced new TSI for SF as an application procedure and it has six phases as follows.

4.1 Phase 1. Become aware of system failure relating to the first principle

Owners of issues and problems understand that they are too complicated to understand from one perspective, and the issues they throw up are too complex to tackle with quick fixes.

4.2 Phase 2. Identify stakeholders relating to the second principle

Owners of issues and problems should identify stakeholders relating to the issues or problems from phase 1.

4.3 Phase 3. Creativity: Identify metaphors relating to the third and the creativity phase in the sixth principle

In the creativity phase, the many different possible views of organizations and their problems should be recognized, and managers and analysts should be encouraged to explore them through the use of Morgan's (1986) "images or metaphors," particularly the machine, organism, brain, culture, and coercive system metaphors. The aim is to take the broadest possible critical look at the problem situation but gradually to focus on those aspects currently most crucial to the organization (Jackson, 2006).

In order to understand system failures, we need models and metaphors. Then methodologies are developed depending upon those metaphors. We introduce three system failure models with metaphors (i.e., the third principle).

4.3.1 Simple linear system failure model (Domino metaphor)

The archetype of a simple linear model explains system failure as the linear propagation of a chain of causes and effects (Heinrich et al., 1980). Figure 8 shows the domino metaphor for this model. The underlying principle is that system failure development is deterministic and there must have cause effect links. FTA (IEC 61025 (2006)) and FMEA (IEC 60812 (2006)) are the representative methodologies. They follow backward and forward chain respectively.

Fig. 8. Domino metaphor

4.3.2 Complex linear system failure model (Swiss cheese metaphor)

The archetype of a complex linear model is well known Swiss cheese model (Fig. 9) first proposed by Reason (1997, 2003). The model put the importance on latent as well as manifested causes. The authors proposed FFSM (Nakamura & Kijima, 2008a, 2009b) as surfacing hidden (latent) factors to suppress deviations leading to system failures.

Fig. 9. Swiss cheese metaphor

4.3.3 Non linear or systemic model (Unrocking boat metaphor)

Perrow (1999) argues that the conventional engineering approach to ensure safety - building in more warnings and safeguards - fails because system complexity makes failures inevitable. This indicates that we need a new model that can manage the system failure. Reason (1997, 2003) explains the organizational life span between protection and catastrophe. The lifespan of a hypothetical organization through production-protection space (Fig. 10) explains why organizational accidents repeat, with this history ending in catastrophe. This is why the periodic application of the methodology prolong system life cycle.

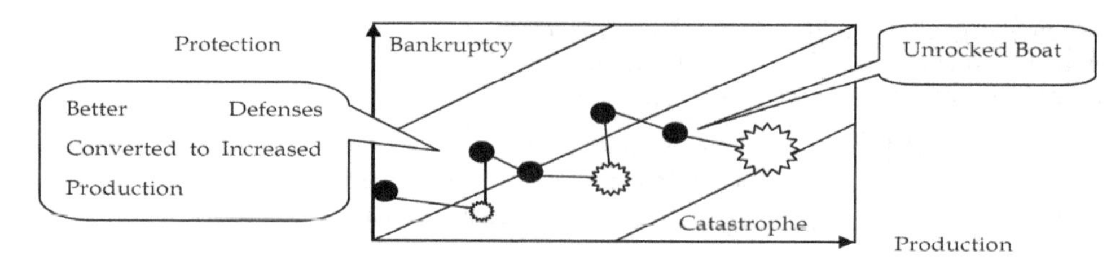

Fig. 10. Lifespan of a hypothetical organization through production-protection space

4.4 Phase 4. Choice: Select methodology using SOSF meta-methodology relating to the fourth and the choice phase in the sixth principle

In this phase, the metaphors generated in the creativity phase are mapped to the SOSF space (Nakamura & Kijima, 2009a) to match the methodology to the problem situation. In the SOSF meta-methodology, problem situations are mapped using three axes (simple/complex, unitary/plural, and Class 1/2/3) in accordance with the degree of (dis)agreement between participants. Problem situations are then mapped to the methodologies as outlined in Table 5. Note that the SOSF meta-methodology is used not to deterministically prescribe which methodology to choose but to illuminate and inform that choice (i.e., the fourth principle).

Model: Metaphor	SOSM Domain	Management Principle	Methodology	Meta-Methodology
Sequential model: Domino Metaphor (Heinrich et al., 1989)	Simple; Unitary	Eliminate Errors	FTA (IEC61025), FMEA (IEC60812)	SOSF (Nakamura, Kijima, 2009ab)
Epidemiological Model: Swiss Cheese Metaphor (Reason, 1997, 2004)	Unitary	Identify Deviations	FFSM (Nakamura and Kijima, 2008a, 2009b)	
Systemic Model; Unrocking Boat Metaphor (Reason, 1997) Rasmussen's Gradients Model (1997)	Plural	Balance Variability	SFDM (Nakamura and Kijima, 2008b), Six Stages (Turner, 1997)	

Table 5. Three system-failure models and their approach to management

We introduce a matrix that clarifies the differences in opinion among stakeholders. Using it helps to clarify the stakeholder views and to identify stakeholders with opposing views. In the example stakeholder matrix in Fig. 11, stakeholders "a" and "b" have opposing views, as shown on the left. After they discuss and debate their views, stakeholder "a" takes responsibility, as shown on the right. In short, a diagonal matrix is created from a non-diagonal one. Table 5 summarizes the system failure models and related methodologies as well as the meta-methodology.

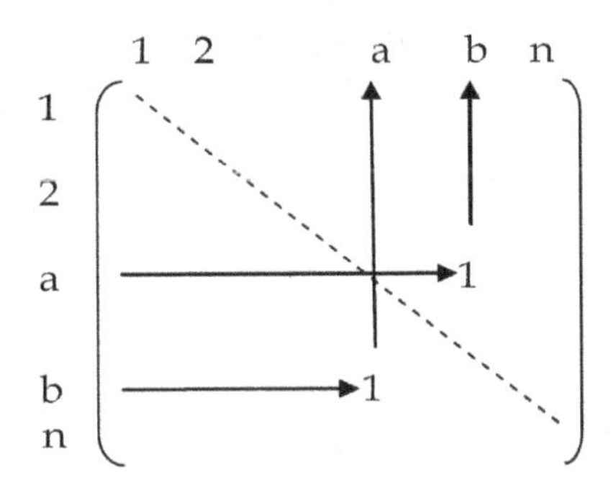

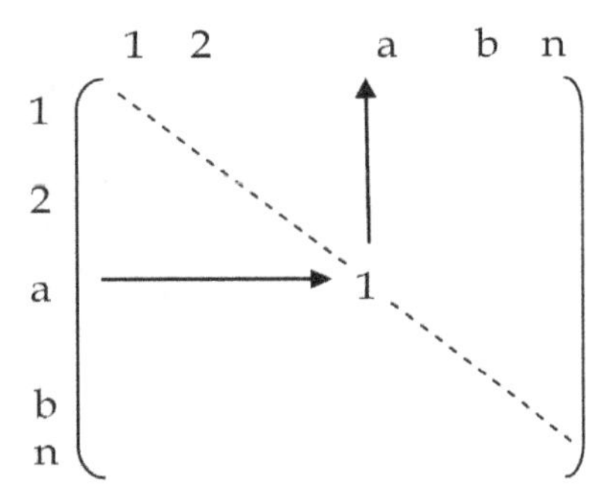

Fig. 11. Stakeholder matrix

4.5 Phase 5. Implementation: Take action relating to fifth and the implementation phase in the sixth principle

In the implementation phase, methodologies are applied to produce change. The methodologies should be used in a complementary way to highlight and address different aspects of organizations and their issues and problems (i.e., the fifth principle). In this phase, the selected methodology in table 5 could be used in accordance with the complementary principles of TSI.

4.6 Phase 6. Reflection: Acquire new learning relating to the reflection phase in the sixth principle

In the reflection phase, the intervention should be evaluated and learning about the problem situation, the meta-methodology itself, the generic system methodologies, and the specific methods used should be produced. The outcome is research findings that are used, for example, as feedback for improving earlier stages of the meta-methodology (i.e., Fig. 12). The relationship between the stages is shown in Fig. 12. There are two feedback loops in Fig. 12. One is to the metaphors (phase3) and the other is to the methodologies (phase4).

5. Application to ICT systems

This section discusses an example application of the TSI for SF methodology to an ICT system failure caused by an operator error resulting from a misunderstanding of the product specifications. In this case, the operator or users who use the products in question was responsible for the failure. The incident escalation procedure is shown in Fig. 13. Those users who encounter the problems of the products report the incident to the help desk, and the help desk provides them with a solution. The help desk then identifies the cause of the incident, and, if it was caused by faulty product design, the help desk escalates it to the development section for further investigation. The development section designs new products on the basis of data for the escalated incidents that the help desk believes were due to product defects. This is mainly because the user-related incidents are screened at the help desk so that the development section can concentrate on product-related issues. The development section measures product quality by AFR (Annual Failure Rate) using only the incidents escalated from the help desk, not by ACR (Annual Call Rate) using all the incidents received directly from the users. AFR is introduced to measure a product quality not to measure a system quality. Therefore AFR is a part of ACR. The metric for product quality is the AFR and system quality that includes product quality is the ACR, which are calculated as shown in Fig. 14.

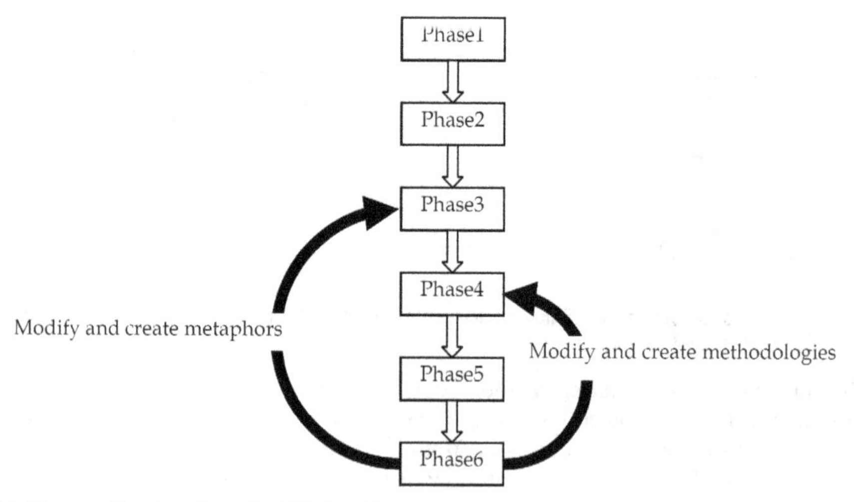

Fig. 12. The application flow for TSI for SF

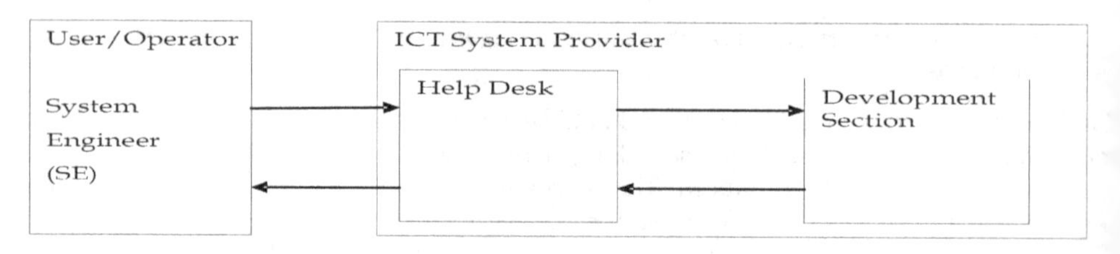

Fig. 13. Incident escalation procedure

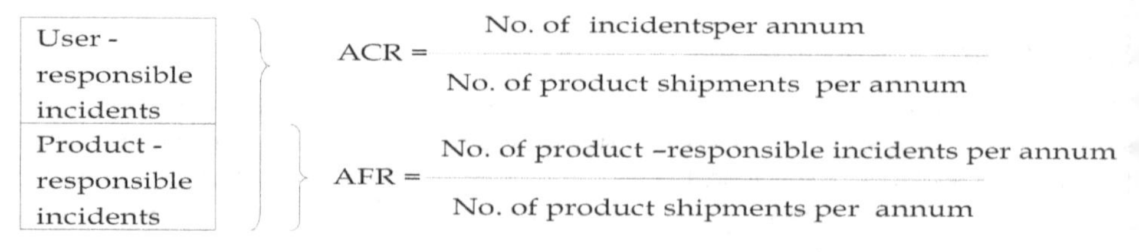

Fig. 14. Calculation of annual failure rate (AFR) and annual call rate (ACR)

As mentioned above, there are six phases in the application procedure for TSI for SF. The followings are the summary of the actual application example.

5.1 Phase 1. Become aware of system failure

In the first stage of intervention, the development section believes that the quality of their product is superior to the average quality of its competitors' products on the basis of internal benchmarking. A third party customer survey reveals that customers judge the quality to be less than that revealed by the internal benchmarking. Upon learning of this discrepancy, the system quality assurance (SQA) section of the ICT system provider sets up a working group to identify the problems.

5.2 Phase 2. Identify stakeholders

The owner of the working group, the SQA section, identifies three stakeholders: an SE (representing a user or operator), the help desk representing the first line engineer, and the development section representing the second line engineer.

5.3 Phase 3. Creativity: Identify metaphors

The SQA section identifies the difference in the key performance indicators (KPIs) between the help desk and the development section. The help desk's KPIs are mainly related to the processing speed and the development section is to the AFR. The SQA section recognizes that increasing the speed should not increase the number of incidents escalated to development section. Furthermore, one way to improve the AFR is to close incidents as user responsible incidents (Fig. 13). Obviously, this may not the best way to handle incidents. Therefore, the two sections' KPIs are not user oriented. The SQA section identifies the unrocking boat metaphor (Table 5) as appropriate for this situation (i.e., the organization is drifting through the environment between excessive economic gain and safety).

5.4 Phase 4. Choice: Select methodology using SOSF meta-methodology

The stakeholder opinions are clarified using the stakeholder matrix (Fig. 11) in order to identify stakeholders with opposing views. As shown in Table 6, the SE and development section have opposing views. The Help desk claims that the SE made an error in operation.

	SE	Help Desk	Development Section
SE	–	–	1: Not an operating error. Problem is product related.
Help Desk	1: Not a product-related problem. Problem is user-related resulting from lack of product knowledge.	–	–
Development Section	1: Not a product-related problem. Problem is user-related resulting from lack of product knowledge.	–	–

Table 6. Stakeholder matrix

The SQA section uses the SFDM to identify three archetypes:

- misunderstanding a Class 2 or 3 failure as a Class 1 failure, (problem)
- erosion of safety goals accompanied by incentive to report fewer incidents (side effect), and
- fix that fails (side effect).

5.4.1 Misunderstanding Class 2 or 3 failure as Class 1 failure (problem)

The source of the failure is inside the help desk system boundary (i.e., a Class 1 failure) although the actual cause is outside the boundary. This archetype (Fig. 15) explains why system failures reoccur following a quick fix or an inappropriate fix. Such fixes might reduce the number of system failures in the short term, but the effects of such fixes gradually become saturated at a level below the organization's goal (i.e., target) level. The balancing intended consequence (BIC) loop becomes open, so quick fixes have no further effect. The balancing unintended consequence (BUC) loop also becomes open as a result of misunderstanding the system failure class and not introducing an effective solution. The sequence of this archetype is from (1) to (5) in Fig. 15. Arrow (1) with the "+" sign indicates that an increase in the number of Class 1 failures causes an increase in the number of actions. Arrow (2) with the "+" sign indicates that the increase in the number of actions increases the number of quick fixes. Arrow (3) indicates that the increase in the number of quick fixes contributes slightly to reducing the number of Class 1 failures. The root cause is outside the system boundary and is unaffected by arrow (4). Therefore, arrow (5) with the "+" sign indicates that the root cause increases the number of Class 1 failures.

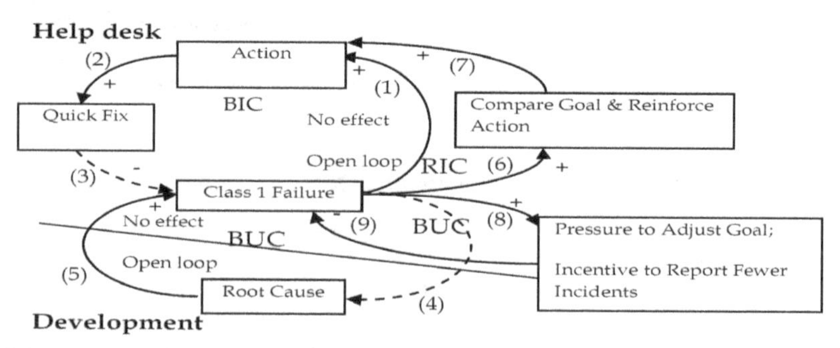

Fig. 15. Misunderstanding system failure archetype

The archetype shown in Fig. 15 is a single-loop learning scenario—a reinforcing action is introduced that is based on the deviation from a predetermined goal. The reinforcing intended consequences (RICs) action to improve the situation leads to the introduction of additional quick fixes, which simply leads to the repetition of a similar scenario. The sequence of this archetype is from (6) to (7) in Fig. 15. Arrow (6) with the "+" sign indicates that an increase in the number of Class 1 failures reinforces the compare goal and reinforce action. Arrow (7) with the "+" sign indicates that reinforcing the compare goal and adjust action increases the number of actions. The RICs action causes various side effects, including erosion of safety goals accompanied by an incentive to report fewer incidents. These side effects are hard to detect because the performance malfunction alarm is muted, and management can identify these effects only by quantitatively measuring performance. This explains why a single-loop learning solution for improving system performance is bound to fail, as Van Gigch (1991) pointed out. In this situation, the root cause outside the system boundary must also be addressed.

5.4.2 Erosion of safety goals accompanied by incentive to report fewer incidents

This side effect is introduced when the RICs loop becomes tighter without a further reduction in the number of system failures (Fig. 15). Increased pressure to achieve the goal emerges from the BUC loop in the form of shifting the goal (i.e., lowering it) and/or hiding the actual state of quality or safety from management. In this relative achievement scenario, a manager who stays within the system boundary has difficulty detecting the actual state of achievement. This is why many Japanese manufacturers have the slogan "3R-ism," which reminds managers to identify a problem at a "real site," confirm it with "real objects," and discuss it with a "real person in charge" before taking any action. The sequence of this archetype is from (8) to (9) in Fig. 15. Arrow (8) with the "+" sign indicates that an increase in the number of Class 1 failures causes pressure to adjust the goal or creates an incentive to report fewer incidents. Arrow (9) with the "−" sign indicates an increase in the number of Class 1 failures that are hidden.

5.4.3 Fix that fails archetype (side effect)

The source of the failure is outside the help desk's system boundary. Figure 16 illustrates a typical example of local optimization. The action taken for the root cause is a short-term solution to the problem that introduces delayed, unintended consequences outside the system boundary, resulting in a Class 2 or 3 failure. For example, an operations manager might shift resources from a proactive task team to a reactive task team because of a rapid increase in system failures, which would only cause the reinforcing unintended consequence (RUC) loop to further increase the number of system failures. This out-of-control situation can only be managed at the expense of others and damages the organization in the long term. The sequence of this archetype is from (1) to (6) in Fig. 16. Arrow (1) with the "+" sign indicates that an increase in the number of Class 2 or 3 failures increases the number of actions within the system boundary. These actions do not attack the root cause (i.e., dotted arrow (5)). Therefore, arrow (2) with the "+" sign has no effect on reducing the number of Class 2 or 3 failures. Alternatively, the arrow with the time-delay symbol (=) might increase the number of Class 2 or 3 failures because of local optimization side effects. Arrows (3) and (4) with the "+" sign introduce an adjust goal and reinforce action without further reducing the number of Class 2 or 3 failures. Arrows (5) and (6) are not in effect during this phase of the archetype.

In this application example, as a result the stakeholders reached the broader and holistic understanding using the SOSF meta-methodology. At initial stage (i.e., preceding stage 5), the user thought these errors are not operation-related but product-related. Conversely, the development section thought they are operation-related. Therefore, the user insisted that they are Class 3 failures of evolution in complex and plural domains in SOSF. Conversely, the development section insisted that they are Class 1 failures of behavior in a simple and unitary domain. Figure 17 illustrates the SOSF space showing all stakeholder opinions.

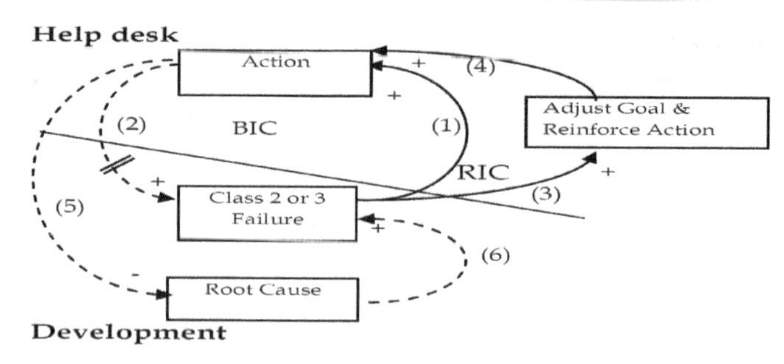

Fig. 16. Fix that fails archetype (side effect)

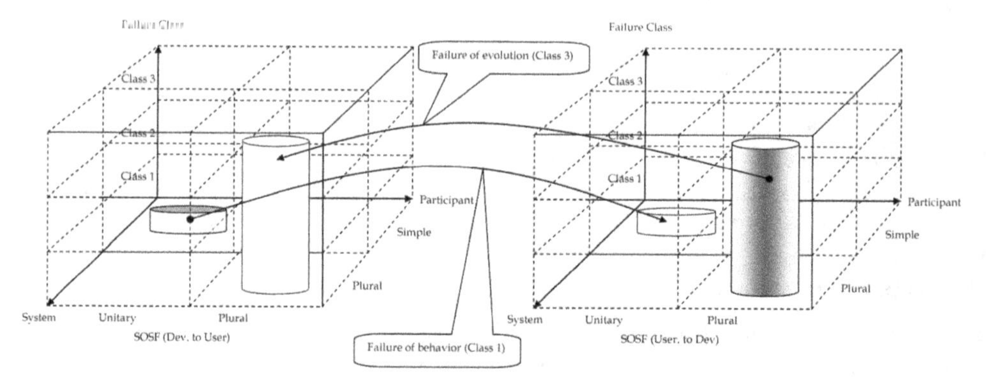

Fig. 17. Simple-Unitary (Class 1) vs. Complex-Plural (Class 3)

5.5 Phase 5. Implementation: Take action

After the debate and discussion, the stakeholders reached the conclusion shown in Table 7.

	SE	Help Desk	Development Section
SE	–	–	–
Help Desk	–	1: It is valuable to expand KPI from AFR to ACR.	–
Development Section	–	–	1: It is valuable to expand KPI from AFR to ACR.

Table 7. Clarify stakeholder opinions using matrix

The SQA section analyzed the user-related incidents and, as illustrated in Fig. 18, judged that 36% of them were possibly product-related. Following their debate and discussion, the SQA section, the help desk, and the development section agreed to change their KPI from the AFR to ACR. The incident reduction scheme is illustrated in Fig. 18. Over the two years of the operation with the new KPI, the ACR have been reduced respectively by approximately 52, 17, 51, and 19% for products A, B, C, and D with the overall average of 36% reduction in Fig. 19.

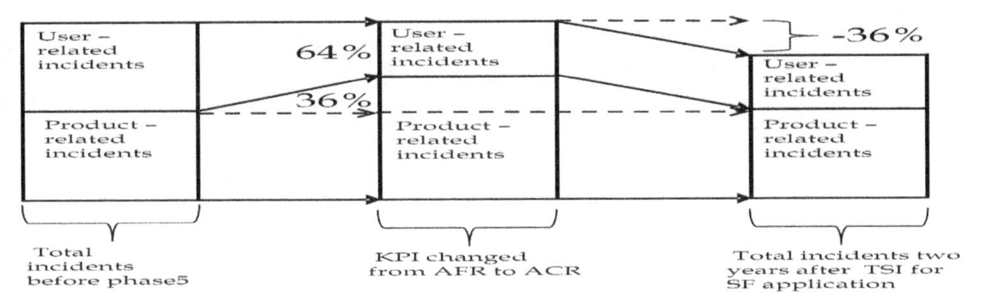

Fig. 18. Incidents transition over two-year period

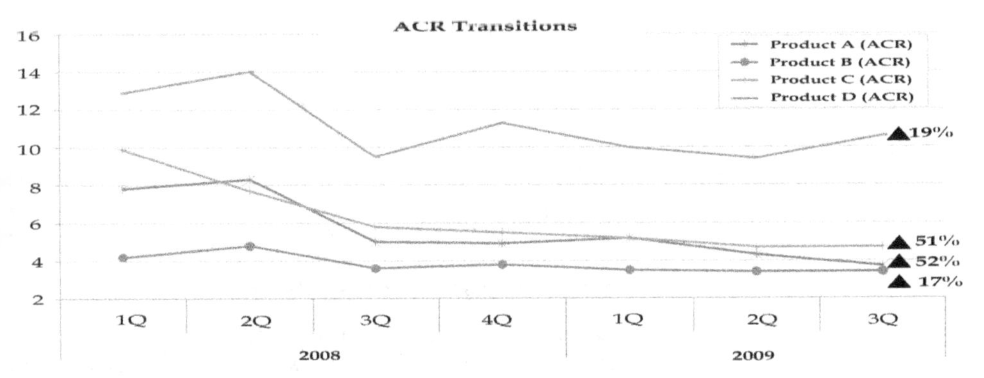

Fig. 19. ACR transitions

5.6 Phase 6. Reflection: Acquire new learning

On the basis of the application example described above, we can identify three ways to overcome the problem of misunderstanding a Class 2 or 3 failure as a Class 1 failure: introduce an absolute goal, close the gap between stakeholders, and enlarge the system boundary. All three actions promote double-loop learning because they alter the process design to improve system quality or safety. In contrast, single-loop learning leads to side effects, as explained for phase four:

- erosion of safety goals and creation of incentive to report fewer incidents, and
- failure of a previous fix.

There are three double-loop learning archetypes.

5.6.1 Double-loop learning for Class 2 failure archetype (solution)

As noted above, it is necessary to focus on the possibilities of relative achievement or the side effects of a quick fix. A tacit assumption of a gap between stakeholders should be surfaced throughout the discussion and debate to close the responsibility gap. Application of this solution to the scenario shown in Fig. 15, misunderstanding system failure archetype, is illustrated in Fig. 20. The sequence of this archetype is from (1) to (6). Arrow (1) with the "+" sign indicates that an increase in the number of Class 2 failures increases the number of actions within the system boundary. These actions induce various side effects (erosion of safety goals or reporting fewer incidents), as discussed above. Arrow (2) with the "+" sign indicates reviewing the stakeholders' mental model gap and redefining or adjusting the ultimate goal. Arrow (3) with the "+" sign indicates provoking a new action. Arrow (4) with the "−" sign indicates that the new action attacks the root cause, which resides outside the system boundary. Arrow (5) with the "+" sign indicates eventually reducing the number of Class 2 failures. Arrow (6) with the "+" sign indicates the path to adjusting the goal and defining the ultimate solution.

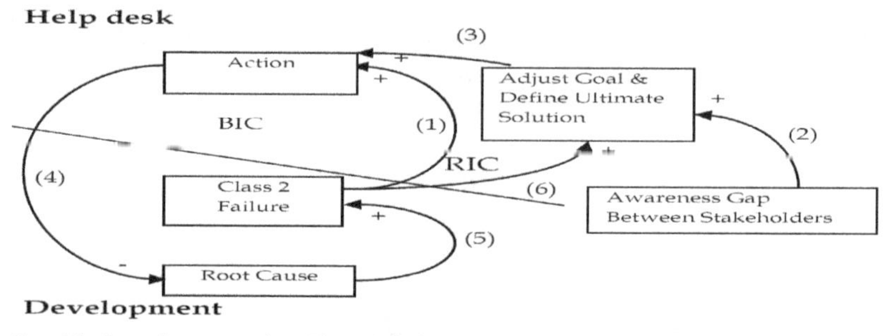

Fig. 20. Double-loop learning for Class 2 failure (solution)

5.6.2 Double-loop learning for class 3 failure archetype (solution)

As mentioned in the introduction, the speed of technology advancement and the growth of complexity are unpredictable. Therefore, a current goal could later become obsolete. This could be the root cause of a system failure, with no party responsible for the failure. In other words, the system failure emerges through no one's fault. This kind of failure can be avoided by periodically monitoring goal achievement and benchmarking competitors. The sequence of this archetype is from (1) to (8) in Fig. 21. Arrow (1) with the "+" sign indicates that an increase in the number of Class 3 failures increases the number of actions within the system boundary. These actions do not attack the root cause, so there is no effect on reducing the number of Class 3 failures, as indicated by arrow (2). Arrows (3) and (4) with "+" signs indicate introducing the ideal goal, provoking awareness of the gap between the current and ideal Goals, and adjusting the goal and defining the ultimate solution. Arrow (5) with the "+" sign indicates introducing a new action, and arrow (6) with the "−" sign indicates attacking the root cause, which reduces the number of Class 3 failures, as arrow (7) indicates. Arrow (8) with the "+" sign indicates further enhancement of adjust goal and define ultimate solution.

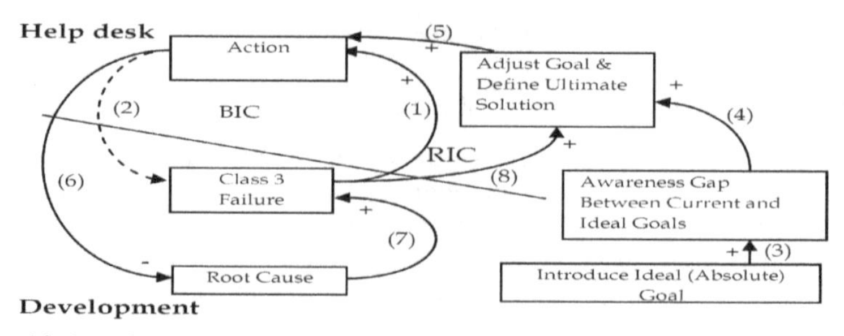

Fig. 21. Double-loop learning for Class 3 failure (solution).

5.6.3 Double-loop learning for fix that fails archetype (solution)

The solution for this archetype is to raise the viewpoint of the problem (Fig. 22). Class 2 and 3 failures become Class 1 if the presumed system boundary is enlarged. The sequence of this archetype is from (5) to (7) in Fig. 22. Arrow (5) indicates enlarging the system boundary to incorporate the root cause. This converts Class 2 and 3 failures into Class 1 failures. Arrow (6) with the "–" sign indicates attacking of the root cause, which reduces the number of Class 2 or 3 failures, as indicated by arrow (7).

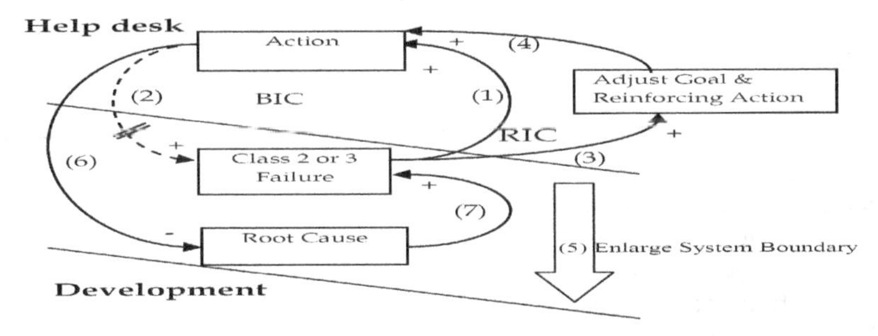

Fig. 22. Double-loop learning for fix that fails archetype (solution)

Figure 23 summarizes the result of SFDM from problem archetype to solution archetype. It shows introducing quick fix (reinforcing current action) is only causing various effects (Erosion of safety goals; incentive for reporting fewer incidents and Fix that fails archetypes).

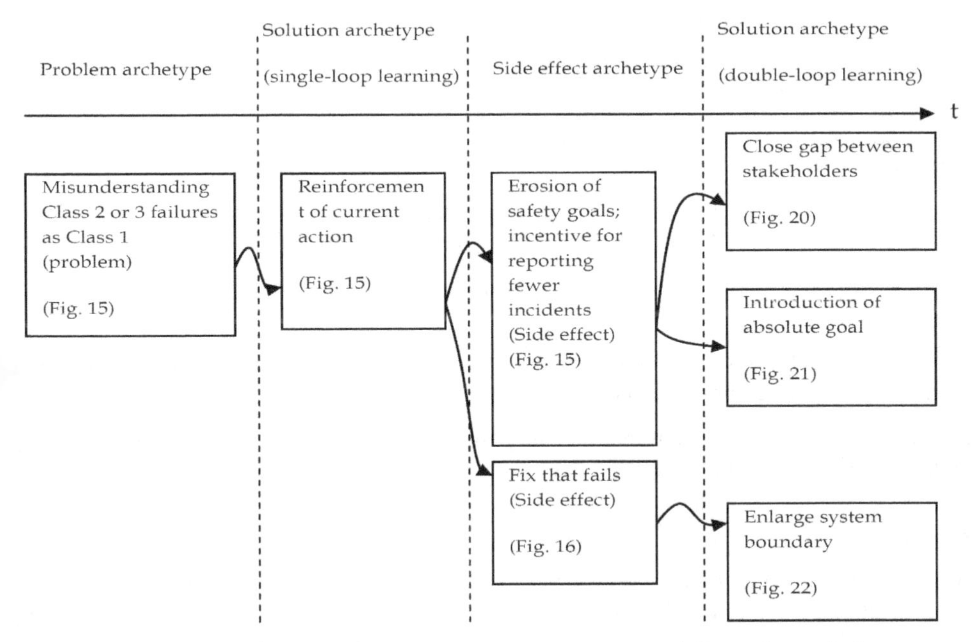

Fig. 23. Problem and solution archetypes in engineering system failures through time.

6. Conclusion

In the ICT engineering arena, the predominant methodologies for promoting system quality and safety are deeply rooted in hard systems thinking. Most organizational processes are reductionist approach. This is reasonable to some extent. Engineers in the development section see systems as the combination of components. The quality of these components determines the quality of the system if the system boundary is defined within the aggregation of components. Therefore, the key performance indicators they use for daily routine processes are not drawn from outside the defined system. In the hard systems thinking paradigm, an efficient approach is to identify deviances from the internal goals and rectify them. The predominant techniques and methodologies play a major role in the simple unitary domain of the meta-methodology called "system of system failures (SOSF)". However in a complex and pluralistic stakeholder's environment, it is clear that several side effects were detected in the "system failure dynamic model (SFDM)" process. This is mainly because the discussion and debate is done among different system levels of stakeholders. The third SOSF dimension represents the responsible system class in VSM terminology. The debate between system 1 and system 5 from different stakeholders could introduce unwanted side effects, as explained in section 5. Especially in the case of failure of evolution in pluralistic contexts, representatives of opposing stakeholders should be from system 5. It is particularly effective in critical system practice, even in the ICT engineering arena, to expand the focus to not only 'work; technical interest' but to 'interaction; practical interest'. The "total system intervention for system failure (TSI for SF)" methodology is useful for changing to an absolute goal learning from the gap between stakeholders and enlarging the system boundary.

We conclude with a summary of the checkpoints and corresponding actions.

Checkpoint 1: Is there a recognizable gap between the perceptions of the stakeholders? If not, use the stakeholder matrix to clarify them.

Action1: Close the gap between the stakeholders. The debate should be conducted with the same system level from stakeholders.

Checkpoint 2: Is your KPI related to absolute goal? (i.e., absolute customers) Do your customers know your KPI? If not, assess the viability of introducing absolute goals.

Action2: Introduce absolute goals to avoid local optimization and to ensure that the essential goal is pursued.

Checkpoint 3: Is the system boundary clear? If not, clarify the boundary. If yes, discuss the feasibility and effectiveness of enlarging the boundary.

Action3: Enlarge system boundary. This would enable to reexamine current system boundary and effectiveness of the process. This could be useful to find out side effects.

7. References

Argyris, C. & Schoen, D. (1996). *Organizational Learning II*, Addison Wesley, 0201629836, Mass.

Beer, S. (1979). *The Heart of Enterprise,* John Wiley & Sons, 0471275999, London and New York

Beer, S. (1981). *Brain of the Firm, 2nd edition,* John Wiley & Sons, 0471276871, London and New York

Checkland, P. (1981). *System thinking, system practice,* John Wiley & Sons, 0471279110, UK

Checkland, P. & Holwell, S. (1997). *Information, Systems and Information Systems making sense of the field,* John Wiley & Sons, 0471958204,UK

Flood, R.L. & Jackson, M.C. (1991). *Creative Problem Solving: Total Systems Intervention,* Wiley, 9780471930525, Chichester

Heinrich, H.W.; Petersen, D. & Roos, N. (1980). *Industrial Accident Prevention: A Safety Management Approach. 5th ed,* McGraw-Hill, 0070280614, New York

Jackson, M. C. (2003). *System Thinking - Creative holism for Managers,* John Wiley &Sons, 0470845228, UK

Jackson, M. C. (2006). Creative Holism: A Critical Systems Approach to Complex Problem Situations, *Systems Research and Behavioral Science Vol. 23, Issue 5,* (September/October 2006), pp(647-657)

IEC homepage, 30.04.2011, available from http://www.iec.ch/

IEC 60812 (2006), *Procedure for failure mode and effect analysis (FMEA),* 4.05.2011, available from http://webstore.iec.ch/webstore/webstore.nsf/artnum/035494/

IEC 61025 (2006), *Fault tree analysis (FTA),* 4.05.2011, available from *http:*//webstore.iec.ch/webstore/webstore.nsf/artnum/037347/

ISO homepage, 30.04.2011, available from http://www.iso.org/iso/home.htm/

Kickert, W. J. M. (1980). *Organization of decision-making,* North Holland, 0444854290, Amsterdam

Nakamura, T. & Kijima, K. (2007). Meta system methodology to prevent system failures, *Proceedings of the 51st Annual Meeting of the ISSS,* Tokyo, Aug. 2007

Nakamura, T. & Kijima, K. (2008a). A Methodology for Learning from System Failures and its Application to PC Server Maintenance, *Risk Management* 10.1, 2008, pp(1–31)

Nakamura, T. & Kijima, K. (2008b). Failure of Foresight: Learning from System Failures through Dynamic Model, *Proceedings of the 52nd Annual Meeting of the ISSS,* Madison, Jul. 2008

Nakamura, T. & Kijima, K. (2009a). System of system failures: Meta methodology for IT engineering safety, *Systems Research and Behavioral Science* Vol. 26, Issue 1, January/February 2009, pp(29–47)

Nakamura, T. & Kijima, K. (2009b). A methodology to prolong system lifespan and its application to IT systems. *Proceeding of the 53rd Annual Meeting of the ISSS,* Brisbane, Jul. 2009

Morgan, G. (1986). *Images of Organization,* Sage Publications, 0803928300, California.

Perrow, C. (1999). *Normal Accidents Living with High-Risk Technologies,* Princeton Paperbacks, 0691004129, New York

Rasmussen, J. (1997). Risk Management in a dynamic society: a modeling problem, *Safety Science,* vol. 27, no 2/3, pp(183-213)

Reason, J. (1997). *Managing the risk of organizational accidents,* Ashgate Publishing limited, pp(3-5), 1840141042, UK

Reason, J. & Hobbs, A. (2003). *Managing Maintenance Error: A Practical Guide,* Ashgate Pub Ltd, 9780754615910, UK

Senge, P. (1990). *The Fifth Discipline: The Art and Practice of the Learning Organization, 1st edition,* Doubleday, 9780385260947 New York

Turner, B. A. & Pidgeon, N. F. (1997). *Man-Made Disasters 2nd edition. Butterworth-Heinemann*, 0750620870, UK

The Columbia Accident Investigation Board Report, 30.04.2011 available from http://history.nasa.gov/columbia/CAIB_reportindex.html, chapter 6 pp(130), chapter 8 pp(185)

van Gigch, J. P. (1986). Modeling, Metamodeling, and Taxonomy of System Failures, *IEEE trans. on reliability*, vol. R-35, no. 2, 1986 June, pp(131-136)

van Gigch, J. P. (1991). *System design Modeling and Metamodeling*, Plenum, 0306437406, New York

Leveson, N. (2004). A new accident model for engineering safer systems, *Safety Science*, vol. 42, issue 4, pp(237-270)

Weick, K. E. & Sutcliffe, K. M. (2001). *Managing the Unexpected: Assuring High Performance in an Age of Complexity (J-B US non-Franchise Leadership)*, Jossey-Bass, 0787956279, San Francisco

3

Systems of Systems: Pure, and Applied to Lean Six Sigma

Ben Clegg and Richard Orme
Aston Business School, United Kingdom

1. Introduction

This chapter will briefly introduce the principles of General Systems Thinking (GST) as defined by classic literature on the subject (Ackoff, 1971; Battista, 1977; Bertalanffy, 1968; Boulding, 1956; Churchman, 1968; Waelchi, 1992; Weinburg, 1975). In particular, this chapter will contrast two opposing operational research (OR) views about systems thinking; the 'reductionistic' ('hard') approach and the 'holistic' ('soft systems methodology') approach. This chapter will then focus on the later; holistic soft systems methodology (SSM), which is the most suitable approach for improving human activity systems, rather than hard systems thinking which is more suitable for mechanistic or physical systems.

SSM may be used for such activities as organisational understanding, process improvement, strategy deployment and change implementation - which are all part of Lean and Six Sigma. A specific type of soft systems thinking will be majored upon, known as Process Orientated Holonic (PrOH) Modelling (Clegg, 2007), which will be used to show how holistic systems thinking can be used to improve organisational performance. The main differentiation between this methodology and any other SSM is that it allows for 'emergent properties' and 'hidden properties' of a system of systems to be depicted by using its unique way of defining a system of systems through the dimensions of 'pitch', 'width' and 'length' (from 'pick-up point' to 'drop-off point'); its abstraction and enrichment rules; the use of holons and holarchies, natural language, story boarding and colourful diagrams.

The case study given in this chapter focuses upon contemporary business improvement trends known as 'lean management' and 'six sigma'. Lean management attempts to reduce waste in an organisation, and six sigma improvement attempts to reduce variation in a process outputs. In reality there is a need to do both, particularly in these challenging economic times. Both lean management and six sigma improvement should be systemic, but this is rarely recognised and even less seldom practiced. The combined technique known as 'Lean Six Sigma' (LSS) is emerging as an attempt to fuse the two approaches together. However a clear concise model has not yet been produced (Pepper and Spedding, 2010). Thus, the current challenge is to produce a unified model of lean management and six sigma improvement that is systematic, systemic and holistic which can be used to optimize systems as a whole. The danger of not applying system of systems thinking to lean management and six sigma improvement initiatives is that different levels (or pitches) of thinking (e.g. philosophy, methodology and tools) and their potential overlap will go unrecognized; and thus their potential impact on organizational performance will be reduced.

This chapter will include discussion about both 'pure' GST and system of systems *per se*, particularly the 'soft' variety which will be of interest to philosophically motivated audiences, such as academic researchers. It will also include 'applied' systems thinking, using PrOH Modelling which will interest audiences wishing to use system of systems thinking to improve 'real world' lean-six sigma systems.

2. Philosophical background to systems thinking

Systems thinking has developed into a discipline in its own right with applicability to almost any area because of its generality (Jackson, 1995). In particular, GST has been applied in the area of scientific management (hard systems methodologies) and more recently to the organizational and human elements (soft systems methodologies). In particular, soft systems thinking has been successful because of its capacity to consider complex situations with competing goals, such as efficiency and quality. Consequently, different types of soft systems thinking have arisen with subtly different purposes; for instance Checkland's Soft Systems Modelling (SSM) is used for general problem definition (Checkland, 1988), Viable Systems Modelling (VSM) to 'diagnose' the operational effectiveness of an existing system and propose their redesign (Beer, 1985), and PrOH Modelling, specifically to improve organisational processes (Clegg, 2007).

GST aims to describe systems with an optimal degree of generality, between the highly generalized relationships of mathematics and the specific theories of the specialized disciplines. Thereby different bodies of knowledge can be combined theoretically into "a body of theoretical constructs which discusses the general relationships of the empirical world" (Boulding, 1956). However it is often difficult for modellers to observe systems as a whole thus systems are not objectively 'real' but subjective; being perceived and inferred by individuals representing what man has created to manage aspects of the world through historical and evolutionary adaptation (von Bertalanffy, 1968). This is particularly the case in behavioural and social systems; in these situations GST represents a method of combining knowledge into a 'system of systems' to act as a 'gestalt' entity (German: essence or shape of an entity's complete form) for theoretical construction. Thus GST can be utilized to organize a large body of information in a way that can identify previously unobserved interrelations; however the subjective nature of systems makes it difficult to prove or refute models developed using GST. Because of such subjectivity it is essential to be clear on ones definition of system of systems. Hence the authors define a system by its constituents and contingency; specifically, people machines, critical success factors, inter-relationships, boundary, inputs, outputs, controls, name and environment. Therefore a system of systems is an entity that is defined by its respective constituents and contingencies *and which may contain other systems and may itself be part of a larger system.* (Churchman, 1968; Ackoff, 1995, 2006).

Boulding (1956) highlighted the need to move away from "mechanical models" that rely on simple, mathematical approaches to better understand the functioning of the World. He utilizes the system of systems approach to GST to arrange "theoretical systems in a hierarchy of complexity" (Boulding, 1956 pp. 202) from level 1, representing simple systems and a low level of understanding, to level 9, representing complex systems and a high level of understanding. He argues that most understanding is at 'level 1' (static structure) or 'level

2', (simple dynamic system) with some disciplines attaining 'level 4', (open system or self-managing structure). Social organizations at 'level 8' exhibit their own characteristics in addition to those of their subsystems, levels 1-7. However the characteristics of a social organization cannot be explained by its decomposition into its constituent parts rather its characteristics emerge from their interaction (Ackoff, 1995). Ackoff (1995) develops the 'system of systems' concept by defining the elements of a system and the changes that occur within them. In this context it is reasonable to suggest that over time an organization which implements lean six sigma and achieves continuous and breakthrough improvement is characteristic of Ackoff's (1995) 'ideal seeking system' which will be composed of other purposeful and goal seeking systems. Thus to begin to understand lean six sigma implementation in a social organization one must attempt to understand its sub-systems and any system of which it is itself a sub-system. By definition this requires a system of systems approach.

A system of systems may "be of value in directing theorists towards gaps in existing theoretical models and might even be of value in pointing towards methods of filling them" (Boulding, 1956). For instance, helping to produce a unified model of Lean Six Sigma practice, as presented in this chapter, is a new, useful, instantiation of this system of systems in the field of management - drawing on operations management, quality, organizational behaviour, and change and leadership literature. Therein the system of systems concept has helped to integrate current knowledge from many related disciplines into a unified holistic conceptual model representing a move towards an ideal system of systems for lean six sigma, which in turn should increase impact of lean six sigma practice on organizational performance.

Some will state that the field of systems modelling originated in the hard mathematical based discipline of systems dynamics (Forrester, 1961) where philosophically speaking, hard systems exist objectively and mainly contain tangible things. As such, hard systems can be engineered to achieve an optimal solution through hard systems methodologies (HSMs). Thus knowledge and understanding of the systems contained in the physical world may be achieved through the application of HSMs such as repetitive experimentation and hypothesis testing (Zhang, 2010). While this may be appropriate for technological or natural systems it has limited value for human activity systems. To fill this gap soft systems methodologies (SSMs), which are based on the same GST principles, but have a significantly different methods of enquiry have been developed. The SSM approach maintains there are no objective systems outside our minds rather they are perceived by individuals based on their particular worldview. Consequently human activity systems often have no singular objective due to the differing aims and goals of the participants resulting in pluralism in problem definition, situation improvement and solving. Thus the outcomes of SSM interventions are not optimal; instead they are compromises that can be accommodated by stakeholders (Checkland & Poulter, 2006; Senge, 1990). Further, optimizing individual aspects of a system in isolation can result in the sub-optimization of the system as a whole; this is particularly the case in supply chain improvements (Forrester, 1961; Ackoff, 1995). Thus SSM has two functions one of logical analysis (or 'structurisation') and one of socio-cultural analysis (the 'function') and can be considered as "one resource of philosophy of social science in theory and practice" (Zhang, 2010 *pp.*165). In this context SSM serves the requirements of rigor and relevance advocated by Tushmann & O'Rielly (2007).

The SSM concept is, by its nature pluralistic, based on perception with the specific goal of systemic intervention. This is achieved through a dialectic process relying on the tension between the objectivist modeller or enquirer and subjectivist positions of the systemees; thereby acknowledging that human activity systems contain different perspectives and are therefore implicitly contradictory (Houghton, 2009). Therefore it could be argued that human activity system models are formed from epistemological pluralism; or in other words, SSMs use multiple ways of knowing and "understand phenomena from a meta-theoretical vantage point" (Houghton, 2009). While this argument has some value, it does not alter the fact that pluralism is constructed from existing accepted forms of knowledge creation, each of which has the same requirements for rigor and relevance. In this light, systems research could be considered as conducting simultaneous enquiries on the same phenomenon from different theoretical perspectives. Such an approach is desirable for researching lean thinking and six sigma practice because (i) the topic is multi-dimensional and interventions should be acknowledged (ii) interventions will consist of a number of stages each of which may be better served by different methodologies (iii) utilizing several methods will have the potential to both increase the "richness and variety" of ideas and outcomes and improve the reliability of outcomes through the application of theoretical triangulation (Mingers 2003). While this may result in better understanding of particular situations it poses a problem for the modeller-researcher when trying to produce a systems model, particularly one that engulfs all the system of systems properties. Mingers (2003) builds on the work of Checkland and suggests that the vital aspect of the pluralist approach is the effective management of the balance between the problem content system (in this case LSS practice), intellectual resources system (in this case the GST and system of systems literature) and the intervention technique (in this case PrOH modelling and potentially systems dynamics). Consequently the combination of methods may vary during the intervention, as a pluralist perspective would suggest, and the appropriateness of the methodologies being combined must be considered from an ontological and epistemological perspective. Thus, if HSMs are misapplied to human activity systems a researcher-modeller may experience difficulty in comparing the model with reality as the model outcomes often contradict the worldview in which it was conceived; which can even occur with poor SSM applications (Ledington & Ledington, 1999). Empirical studies into the use of pluralistic approaches include Mingers & Taylor, 1992; Munro & Mingers, 2002 and Mingers & Rosenhead, 2004).

In order to mitigate the difficulties of methodological misapplication 'system of systems methodologies' have been developed (Jackson, 1990). However this approach can create difficulties, as with SSMs above, if assumptions (worldview) used when 'reading' the problematic situation turn out to be inappropriate. A similar approach, Integrative Systems Methodology (ISM), uses a framework of opposing perspectives, (e.g. objective and subjective) in order to develop a tension between them allowing effective management of complexity (Schwaninger, 1997). This is also similar to the concept of creative tension promoted by Senge (1990).

3. Contrast of reductionistic and holistic systems thinking

Agricola (1556) developed a systematic analysis approach to operations management by documenting his scientific and empirical evidence and used it to challenge contemporary theory and practice at the time; this was also adopted by the likes of Charles Babbage (1835) to develop "industrially relevant but conceptually robust" advances from the combination of theory and practice. Subsequently F.W.Taylor (1911) developed the 'Law of Heavy Labouring' and the theory of scientific management which, through the identification of key components of performance, standardized and reduced processes to fundamental levels. Both Taylor and Henry Ford (in 1926, reported in Ford, 2003) separated the planning of work from its execution utilizing experts to develop processes which were then implemented by workers. Subsequently, statistical approaches were developed by W. Shewart and W. Gomberg which formed the basis of mass production. In 1971 Mintzberg (1971) challenged Management Science to expand the reductionist approach beyond processes to develop the understanding of management practices to describe them precisely and understand management as a 'programed system'. These attempts to reduce operations management to fundamental components where improvements can be made through data collection and analysis or experimentation proved successful in the early 20th century but have not continued to provide the same level of insight (Sprague, 2007). This is particularly the case when considering the superior performance of Japanese motor manufacturers, who advocated a 'holistic' systemic approach of lean management, compared with their American counterparts, who advocated reductionist approaches (Liker, 2004; Womack *et al*, 1990).

Interest in 'holistic' systems approaches developed in the 1950/60s as a challenge to the one-way causal paradigm of reductionist approaches (von Bertalanffy, 1968). Proponents argue that the World cannot be understood through the decomposition of systems into their component parts. Such systems consist of 2 or more inter-related elements each of which affects the whole; thus the properties and behaviour of each element and its resulting effect on the whole depends on at least one other element. In this context the whole cannot be understood through the aggregation of reductionist parts because the characteristics of the whole are a result of their interactions (Ackoff, 1974). Additionally the reductionist approach was challenged by Forrester (1961) stating that mathematical optima had little applicability to 'real world' problems due to oversimplification becoming devoid of practical interest and therefore utility. Further still, management as an art was more complex, difficult and challenging than mathematics, physics or engineering because of the greater scope of systems and the numerous non-linear relationships that control the course of events. By 1969 Wickham Skinner (Skinner, 2007) began to question how the application of accepted managerial principles in businesses could results in failure; concluding that the optimization of individual aspects, such as production and marketing, could pull in different directions because of their differing goals. Consequently, it is necessary to fit the components together as a strategy, in order that the system functions as a whole to achieve a specific aim.

There are a wide range of methodologies that could be applied in the field of operations management, but for the purpose of this chapter the focus will be on the systems approach and the dominant quantitative and qualitative methodologies therein. In particular, soft systems approaches which aim to produce models of a 'problem situation' to facilitate better understanding, enabling conclusions to be drawn about appropriate corrective action.

3.1 Systems Dynamics (SD)

Systems Dynamics (SD) was developed by Forrester (1961) at MIT as a methodology designed to produce representative models of the complex patterns of dynamic relationships between the 'stocks and flows' of physical or social processes. Using the concepts of ratios, levels, feedback loops and control, the dynamics are assigned causal and mathematical relations, which can then be used to predict the effect of different types and levels of intervention. Philosophically, "problems can be separated from context, and treated in a purely theoretical way, to pursue objective information to find and display a scientifically demonstrable solution" (adapted from Lane & Oliva, 1998 *pp.* 225). Thus by definition, the models are an ontological description of the problem, they are an objective representation of the pre-existing real world independent of context which can be fully observed by a detached researcher. As such they represent a realist and reductionist perspective. Epistemologically knowledge is created through representation by modelling, which can be used to form dynamic hypotheses of how the 'reference mode' (situation under investigation) causes the observed behaviour. SD is a quantitative methodology using observation and measurement combined with judgement and opinion by an analyst to optimize the system under investigation (Mingers, 2003) using a realist, functionalistic, determinist and 'hard' systems methodology fitting into the area of functionalist sociology (Lane, 1999).

In order to model systems using SD and other hard approaches human activities are simplified in order to allow mathematical description of processes. To do this various assumptions are often made (Boudreau *et al.*, 2003):

- people are not a major factor (OM models focus on machinery, frequently omitting human effects)
- people are deterministic and predictable (people have perfect availability, are identical and uniform, task times are deterministic and mistakes are random or do not occur
- workers are independent, individuals unaffected by others
- workers are stationary, workers do not learn, problem solve or exhibit any patterns of behaviour such as fatigue or motivation
- workers support the production or delivery of the service but as not part of it, the impact of system structure on the interaction of customer and worker is ignored
- workers are emotionless
- work is perfectly observable; measurement error and the Hawthorn effect are ignored.

While this moves beyond the simple cause and effect of the natural sciences criticized by Forrester (*ibid)* the effect is to assume that human behaviour is of little consequence. Reacting to this, the field of behavioral operations has emerged, challenging the simplification of human behavior in operations management modeling and questioning a number of assumptions which form the paradigm of operational research. Therefore, hard systems approaches such as SD are subject to many criticisms:

- there is no method for assessing the appropriateness of the chosen worldview or means by which other worldviews can be articulated (Lane & Oliva, 1998)
- they do not consider power and social interactions therefore human actions are rational and neutral
- does not distinguish the problem solving system from the real world problem (Checkland & Poulter, 2006)
- as they do not give refutable hypotheses (Peery, 1972), nor a method for translating models into real world action. As such, SD provides representation of the situation but no ideal vision (Rodrigues & Bowers, 1996).

3.2 Soft Systems Methodology (SSM)

In contrast to SD, SSM is an interpretive approach; as such the real world is not detached from the model, but constructed by those who experience it; hence a real world system is by definition subjective and context specific to the person experiencing it. SSMs are designed to create models of 'real world problem situations' in the 'systems world', so that the participants can better understand the problem and reach an 'accommodation' - a solution acceptable to stakeholders - on action (Checkland & Poulter, 2006; Checkland and Scholes, 1990). Consequently, SSM does not aim to optimize or solve a problem in the way that SD does, but instead sees problems as 'messy' human processes requiring constant negotiation (White, 2009). In SSM, model building is considered to be a social process, a personal experience that can only be understood as a whole, which produces useful devices that can be utilized to "help human agents to create their social worlds via debate and the construction of shared meaning" (Lane, 1999). As such, the different worldviews of the participants regarding the 'real world problem situation' can be considered (Mingers 2003). Ontologically, SSM identifies a real world problem but treats this in an interpretive fashion, thus the world exists in the context of human activity systems (sometimes known as 'holons') which are perceived according to the worldviews of those involved (Mingers, 2003). As such, stakeholders have their own motivations and goals, and real world problems can only be solved as acceptable compromises (Checkland & Poulter, 2006). Knowledge of the problem situation is created (epistemology) through the use of systems concepts, rich pictures and logical relations. Thus hard and soft information about the problem can be assembled in the context of the worldviews of those concerned; as such the information assembled is predominantly qualitative. The models produced can then be used by analysts, researchers, facilitators and the participants to learn about and improve a real world problem situation by achieving consensus on feasible and desirable changes (Mingers, 2003; 2011).

The main criticisms of SSM is that it is difficult to implement, as it is not a problem solving method but rather a method to enable better consideration of the problem at hand. As such, it is difficult to assess the outcome of the SSM which, as a pure qualitative method, cannot produce a measurable outcome. Criticisms of SSM include:

- modelling on the basis of different worldviews makes problem definition difficult, and the selection of a worldview means the real world problem is not modelled instead producing 'ideals' from a particular worldview (Lane & Oliva, 1998)
- no mechanism by which the conclusions can be implemented as it assumes implementation will happen automatically because it is the logical accommodation of stakeholders. Because the output is not a system, the proposed changes are not necessarily systemic and therefore infeasible (Lane & Oliva, 1998) and may lack cybernetic alignment (Flood & Jackson, 1991).

3.2.1 Process Oriented Holonic (PrOH) Modelling

Process Oriented Holonic (PrOH) Modelling is one of the more recent versions of SSM with philosophical bases in GST and system of systems. The purpose of PrOH modelling is to produce a systemic set of models without using a reductionist or mechanistic approach to modelling. This is necessary because human activity systems have high levels of stochastic, non-determinist and sometimes irrational and illogical behaviour (Balogun & Johnson, 2005; Rice, 2008). To approach this from a reductionist perspective would lose information necessary to see 'hidden' and 'emergent' properties at different organizational levels. As such, PrOH modelling is a methodology effective for linking strategic vision to operational processes. It is best applied to environments of high complexity, low volume and high variety where opportunities for repeated learning are limited; as is typical in LSS implementations in organisations.

PrOH utilizes pictorial models using 'bubbles' and linkages to describe processes. Uniquely PrOH defines the 'scope' (or area of interest) using three dimensions; pitch, width and length. This allows modellers to properly define their models and allow easier validation. The 'pitch' of the model is the organizational level being modelled; it is usually only necessary to use three levels strategic, tactical and operational. The 'width' of the model relates to how much detail of the supporting activities of the core process is included and the 'length' defines the 'pick-up' and 'drop-off' points of the model, in other words, its beginning and end.

Each PrOH model is built around a core process, making validation and the level of granularity easy to define, using bubbles to represent 'nouns' such as people or things which are linked together using arrows, utilizing verbs to describe the connection or linking arrows. The major advantage of this approach is the promotion of natural language, limiting codification, making the models more accessible to users at all levels. Natural language instead of jargon will make systems thinking more accessible and increase practice (Ackoff, 2006).

In summary, PrOH modelling produce 'holarchies' based on abstraction and enrichment suited to complex problems such as implementations of new systems, organisational changes or large high volume and low variety projects. In contrast, hard systems methodologies, such as SD, produce hierarchical models based on aggregation and reduction that are best suited to low variety, high volume relatively short lead time processes such as seen in white goods or automotive manufacturing, where learning opportunities and data collection opportunities are repeatedly available.

While lean thinking and six sigma can utilize specific reductionist tools for their implementation, when considered overall, as a holistic approach, LSS is better suited to GST and system of systems, particularly SSM approaches, which represent the complexities of the processes and organizational change. Therefore the authors recommend that a combined hard-soft approach should be used for modelling LSS projects and programmes.

4. Systems thinking for Lean Six Sigma

Lean thinking and Six Sigma are used to improve unstructured systemic problems, and can help transform organizations when properly deployed; for example, Toyota (Lean thinking), General Electric and AlliedSignal (Six Sigma) are often quoted. Typically organizations have adopted one of these lean or six sigma techniques but, more recently, the combined approach of LSS has emerged as an attempt to fuse the two approaches together in a way that combines their strengths and mitigates their weaknesses. However there is currently no definitive, highly impactful way of achieving this despite high profile successes (Pande, 2000; Kwak, & Anbari, 2006) with tools being widely utilized and accepted by academics and practitioners alike. Therefore, the question must be asked - why do some organizations fail to achieve the results they expect?

Benefits of lean six sigma were initially promoted in management guides by industry gurus (Pande, 2000; George, 2002; Martin, 2006) and subsequently in trade press articles describing successful projects (Anon, 2010; Burgess, 2010). However they are prescribed solutions, a one size fits all solution, which cannot account for observed failures. Thus reasons for differences in deployment success are unclear, stimulating academic research into LSS programmes and their successful implementation. Research approaches utilized fall broadly into five groups (Brady & Allen, 2006):

- case study - often of specific situations with little wider applicability
- comparative - describing relative merits of different techniques
- theoretical with application - modifications to methodology applied to specific situations
- surveys - of application and desirable traits of implementation and sustainability
- literature review - describing gaps in literature, issues and future research.

While the literature provides valuable insight, they are limited in scope and focus on the desirable traits of successes, rather than how the current pool of knowledge might be integrated. Currently there is little rigorous academic examination of the effective implementation of key factors governing the initial and ongoing success of the LSS strategy, and as yet there is no definitive model for its deployment or unified theory with general applicability (Furterer & Elshennawy, 2005; McAdam, & Hallet, 2010; Thomas *et al*, 2009; Tjahjono 2010; Naslund, 2008; Nonthaleerak and Hendry, 2008; Pepper and Spedding, 2010; Proudlove *et al*, 2008). Given that many desirable traits and methodological insights have been identified, there is a need to identify how they can be more effectively integrated and deployed.

Lean thinking and six sigma utilize reductionist tools and statistical analysis applied to processes to improve them, but their success is reliant on the contribution of employees. Consequently, people and teamwork are major factors; therefore it is not logical to assume behaviour is predictable or emotionless (Robbins, 2001). Each technique utilizes workers to problem solve, learn from mistakes and improve processes to deliver what the customer desires. Organizations utilizing these techniques are driven by a focus on quality, efficiency and effectiveness, as defined by their customers and the workforce, as an intrinsic part of this process (Shah & Ward, 2007; Pande, 2000; Womack *et al.* 1990; Clegg *et al.* 2010). Additionally, lean thinking is designed to eliminate the 'hidden factory'; which is present but not often observed by traditional operations management. Thus one could consider lean thinking and six sigma to be modern incarnations of the principles of scientific management in terms of process management, tools and statistics (Chakrabarty and Tan, 2009), but with a significant focus on the human aspects of operations management (Zu *et al*, 2010).

Therefore success of lean thinking and six sigma in practice is not only a result of its scientific origins, but the way in which the workforce is utilized. Therefore approaching issues of implementation and management of lean and six sigma programmes from a purely scientific and reductionist viewpoint cannot address important issues such as the systemic functioning of implementations (Naslund, 2008; Conti, 2010; Kuei & Madu, 2003; Baines *et al*, 2006). For example Clegg *et al.* (2010) conclude the effectiveness of any aspect of lean and six sigma implementations is the product of technical quality and cultural acceptance, not just the summation of reductionist parts. Similarly the emerging field of behavioural operations attempts to merge the perspectives of HRM and OM in order to understand the people aspects of operations management (Boudreau, 2003; Linderman *et al.*, 2006; Bendoly & Hur, 2007).

4.1 Criticisms of LSS

While there is a significant quantity of rigorous empirical evidence to guide the use of LSS, the current methodology does not account for the human side of implementations (Chakravorty, 2009). Indeed widening the concept and application of LSS, to include people and organizational criteria will aid embedding (McAdam & Lafferty, 2004). Additionally organizations must balance flexibility and people-oriented cultural values with the need to maintain control systems (Zu *et al*, 2010) as often deemed critical, even in innovative processes (McComb *et al.*, 2007; Naveh, 2007; Jayawana & Holt, 2009); to achieve this, the organization must continuously adapt (Kwak & Anbari, 2006).

It has been argued that lean thinking, six sigma and LSS are just repackaged versions of earlier techniques, and are all essentially fads (Naslund, 2008). A primary criticism is the isolated nature of lean and six sigma leading to problems of compartmentalization, sub-optimization and fragmentation causing the organization to suffer as a whole (Naslund, 2008). This isolation can result in the benefits of training not being realised (Lu & Betts, 2011). Similarly, Conti (2010) states that strategic fragmentation results from a lack of systemic perspective and there is a joint role for quality and systems thinking in value generation. Thus "quality is not a result of one factor or issue but is rather systemic" (Kuei & Madu, 2003); as current management foci, social and technical system components must be understood and managed through systemic implementation. Further, Naslund suggests "the theories behind systems thinking applied via process management can provide the framework needed to facilitate and maintain successful organizational developments". Thus "a truly successful application of lean requires organization-wide changes in systems practices and behaviour (Baines *et al*, 2006). Other researchers identified a need for a systemic approach to achieve the best results in LSS (Proudlove *et al*, 2008) and appropriate selection of the most rewarding projects (Su & Chou, 2008; Yang & Hsieh, 2009). Additionally, the application of a systemic view has been shown to increase deployment success through organizational learning (Wiklund & Wiklund, 2002). Such learning is vital for an organization to adapt to the needs of the system in which it operates; in this respect prescribed solutions are too simplistic and often fail to account for hidden dynamic systemic issues, which can help to understand potential, distant, time delayed side effects of improvement actions (Senge, 1990). This has led to calls for future research into systems, to combine current knowledge and allow the dynamics of implementation to be monitored, understood and optimized (Brady & Allen, 2006, *pp*.348).

4.2 Towards a solution

Given that the adoption of LSS is an organizational change, it is logical to use tools that are flexible enough to accommodate the dynamics of the process, whilst also providing a way of visualizing the desirable traits and making them accessible to everyone. The systems discipline argues that there is no such thing as an isolated process, that all processes are linked through dynamic interactions that must be considered as such when approaching organizational problems. Systems thinking is a methodological body for studying and managing complex feedback systems such as one finds in business and other social systems. It is an approach in which the model resembles reality structurally so it can be reviewed for usefulness and consistency. Advocates of the method state "until one understands the dynamic cause of present undesirable conditions one is not prepared to explore moving from present conditions to more desirable conditions" (Forrester, 1994). This can be achieved by either of two approaches. Firstly, mapping the dynamic relationships then using a variety of methods to understand the possible consequences of those relationships or develop theory. Secondly, creating a model of the dynamic relationships in a given problem or situation in order to explore the consequence of different amounts of intervention, timing delay and feedback. It is important however to realise that it is not designed to give the right answer but to help consider the problem more effectively (Forrester, 1961; Checkland & Poulter, 2006).

Although the importance of a systemic view is highlighted by authors there is no unified systemic and systematic approach, which is of vital importance for both the successful implementation and maintenance of LSS initiatives. Such an approach could crystallize the knowledge of previous researchers and help to mitigate or eliminate the problems and issues they have identified. Such work could allow the development of a LSS organization rather than an organization carrying out LSS projects.

Unified systemic models may provide a new perspective to the combination of lean and six sigma in the context of systems thinking. In the academic arena this will bring together the previous knowledge and provide an integrated LSS methodology and theoretical model for the deployment of LSS called for in contemporary literature. Additionally applying systems thinking to the LSS field will enhance the ability of practitioners to understand the processes they are trying to change or improve. Superior results will be derived from systemic insight, going beyond quantitative measures inherent to LSS to examine the soft aspects of organizational culture. The ultimate goal of a systemic LSS model would be to produce an organizational structure that is optimized within the LSS framework.

5. A proposed system of systems model for Lean Six Sigma

The overall aim of both Lean Thinking and Six Sigma is to bring about improvements to processes that will benefit the organization as a whole. In order to realize this aim the organization must first assess the needs of the current business environment and translate them into a clearly defined strategy, objectives and goals to form the basis of organizational activities. A LSS Champion should be responsible for promoting the LSS programme and acts as a link between the Executive and the wider organization. As such a LSS Champion considers strategic objectives and goals and develops LSS Projects appropriate to their achievement. These projects are then assigned to and managed by LSS Black Belts; subsequent process improvements are maintained by Process Owners improving the organizations position in a Future Business Environment. These roles and entities (shown as proper nouns in this paragraph) are defined in Figure 1 as the core process for LSS involves development of LSS projects based on the strategic requirements of the business.

The nature and language of PrOH models utilizes the use of natural language to make the models easier to follow reducing the need for the reader to need specialist training and ensuring accessibility for any user. Thus the core process for the strategic level of LSS deployment is shown in the first PrOH model given in Figure 1 which reads as *'The Current Business Environment may require the Executive Board to develop a Strategic Action Plan to inform the Lean Six Sigma Champion who develops Strategic Lean Six Sigma projects which are managed by the Lean Six Sigma Black Belt to improve the Future Business Environment.'* Also, the Executive Board consists of the CEO and representatives from Operations and Finance, similarly the Lean Six Sigma Black Belt is responsible for the management of projects, but they are supported by the Process Owner for the process under consideration in the project. The Strategic Action Plan is shown to focus on growth, acquisition and cost. Through further enrichment at the lower levels, tactical and operational, the nature of these aspects of strategy could be further refined, providing an enriched set of models in a holarchy describing LSS deployment, which have produced using a soft systems approach.

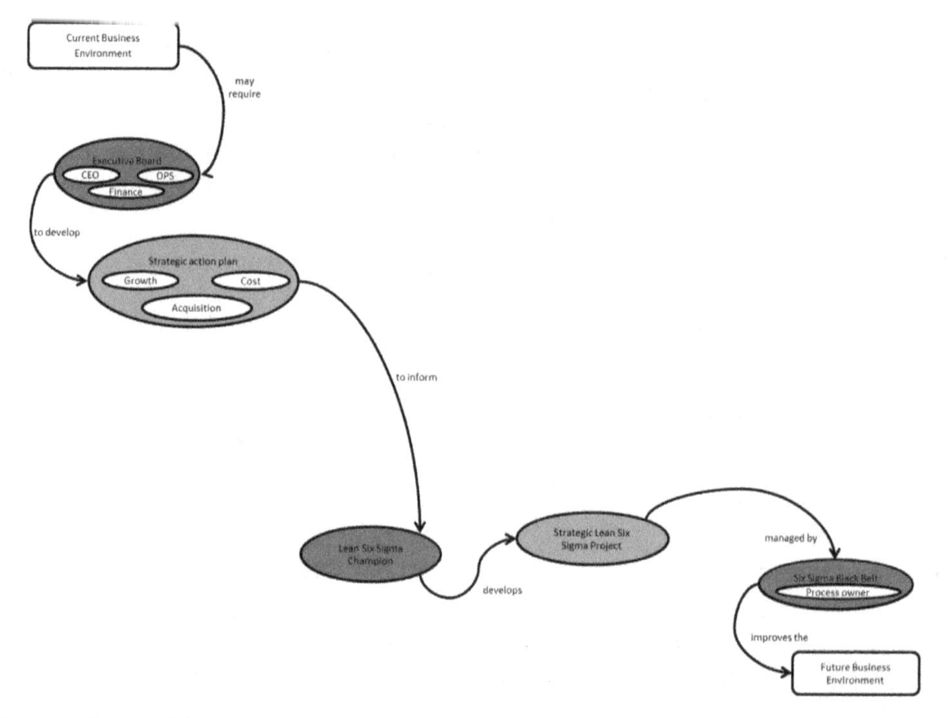

Fig. 1. PrOH Model - Core Process

In order for the core process to be executed effectively it is necessary to have supporting activities to facilitate it, these can be added to the model as in Figure 2. So the development of the *'Strategic Action Plan directs the Departmental Managers to produce Goods and Services which are monitored by Performance Managers who produce Reports to inform the Lean Six Sigma Champion'* who utilize it to develop Strategic LSS Projects. Additionally the reports on the current performance of departments will be used by the LSS Black Belt as part of management of the project. These supporting processes could also be enriched in the same way as the core process. An additional set of supporting processes would cover external aspects such as market data. Thus *'the Strategic Action Plan informs the Customer Relations Manager who collects Customer and Market Data to inform the LSS Champion.*

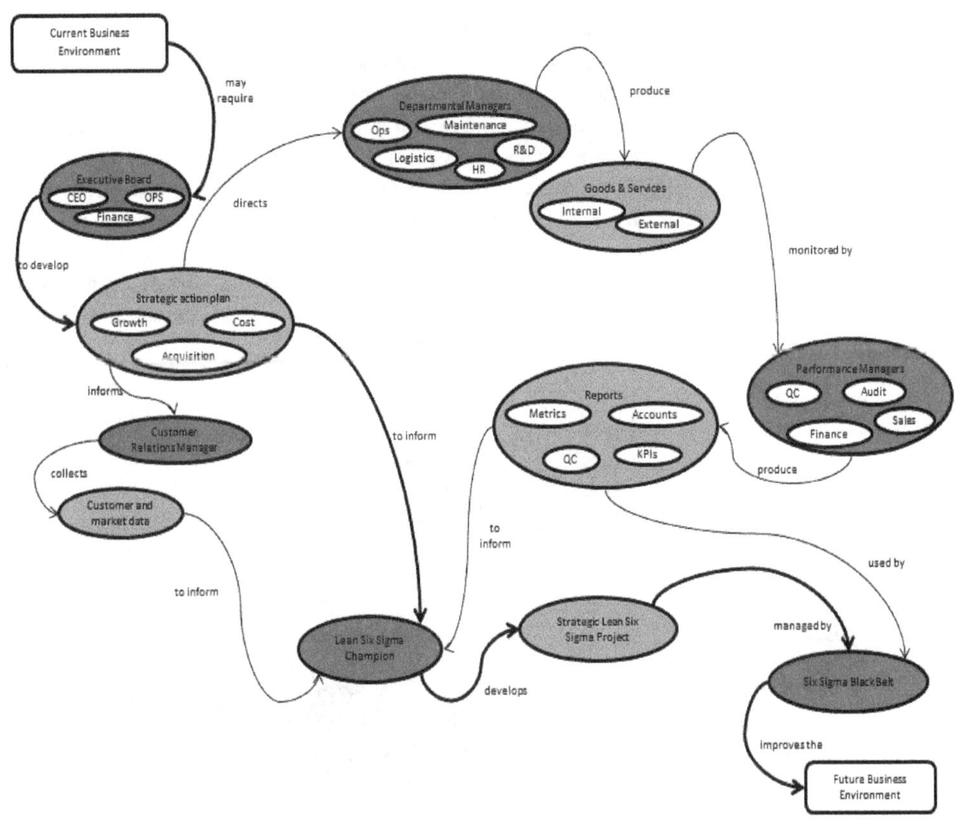

Fig. 2. PrOH Model Showing Supporting Processes for the Core Process

In addition to the strategic initiative to conduct projects; problems may arise in the day-to-day business operations that need to be addressed by LSS projects. One possibility is a change in market conditions such as the emergence of new competitors, market opportunities or changes in customer requirements which require an urgent organizational response. Another possibility is the emergence of problems in the day-to-day operations. Each various department manager will have been assigned objectives and goals through the strategic action plan which have directed them to produce goods and services for their respective internal and external customers. The performance of these departments is then measured and reported against targets by various performance managers/departments, such as quality control and finance. Should these reports identify failures in the organizational performance it may be necessary to conduct LSS projects to rectify them. Additionally this information is vital to the LSS Champion when deciding which potential projects should be given priority and the LSS Black Belt when executing LSS projects.

The 'holistic' lean thinking approach utilizes the 'Deming cycle' of plan-do-check-act (PDCA) to drive continual improvement. The cycle iterates between *planning* and *deploying* long and short term organizational objectives and goals. *Doing* is completing operational activities to achieve those objectives and goals. '*Checking*' *is* the performance of those activities and *Acting* to improve the *Doing* of those activities leading to a further cycle of plan-do-check-act. *Acting* may be specific projects designed to achieve long and short term goals or ad-hoc improvement actions to address local problems that emerge in day-to-day operations. However lean thinking does not have an explicit mechanism by which projects should be executed with improvement actions being carried out 'as required' and not necessarily benefiting the organization as a whole which is overcome by the incorporation of six sigma into the Act part of the cycle, giving a structured approach to the execution of continuous and breakthrough improvement. Conversely, the addition of the Plan-Do-Check to the formal project approach of six sigma will provide links to organizational strategy and the performance of operational processes necessary to select impactful projects, which is currently missing in traditional six sigma implementations.

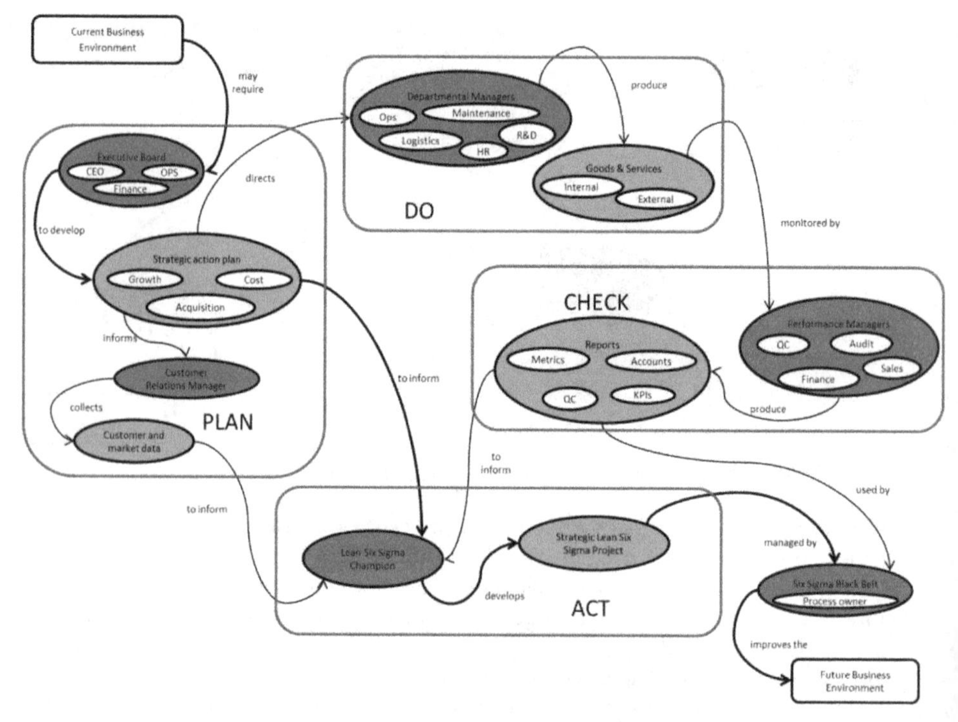

Fig. 3. The Deming Cycle Applied to the Lean Six Sigma PrOH Model

Thus it is beneficial to perceive lean six sigma as described in this chapter - because defining project improvement structures in six sigma decides the method by which the Act part of the lean thinking cycle is achieved and the remaining iterative part of the Plan-Do-Check cycle provides necessary structure for the selection and development of appropriate LSS projects that lead to the achievement of tactical and strategic objectives. Figure 3 shows Deming's Plan-Do-Check-Act cycle imposed upon the Lean Six Sigma PrOH model. The reason for doing this is to show that a SSM such as PrOH can be used to describe the LSS paradigm, whilst a more quantitative meta-system (such as the Deming cycle) can be super-imposed on it (shown by large blue boxes), in which systems dynamic performance measures can be applied. Such metrics should be based on the costs of quality - failure, prevention and appraisal (Kiani, *et al.*, 2009, Pavlov and Bourne, 2011). The implications for managers is that they need to ensure that the activities of an organisation, which can be shown in enriched PrOH models at strategic, tactical and operational levels need to be aligned with metrics in the meta (PDCA) measurement system.

6. Conclusion

Systems approaches can provide representative models of the real world and move beyond the general simplistic linear cause-and-effect relationships which are appropriate to physical systems. Indeed such an approach is necessary if human understanding of the world is to move beyond the superficial, simplistic mechanistic models (Boulding, 1956). By utilizing systems thinking and drawing on a range of literature, the possibility of creating holistic understanding, unconstrained by ontological or epistemological tradition, whilst maintaining academic rigor, may be achieved. Additionally, correctly applied, systems approaches can produce models that are more relevant to praxis; resulting in methods and approaches that can facilitate desirable changes in organizational policy deployment (Akao, 1991) rather than abstract, theoretical, 'academic' solutions.

While PrOH modelling provides a valuable tool for the modelling and discussion of the structure of LSS processes as applied in this chapter, it is limited to qualitative assessment of human activity systems. PrOH is of value in the development and understanding of the unification of lean and six sigma as it can provide an ideal view of LSS. However, it does not provide a method by which the appropriateness of the conclusion can be quantitatively assessed. Thus the model as in Figure 3 may be defensible but it does not provide a method to assess its quantitative ability to produce the desired outcome. By combining PrOH modelling with a meta-measuring system based on Deming's Plan-Do-Check-Act cycle using cost of quality metrics in an SD model it should be possible to produce models that clearly articulate LSS processes (PrOH) and a method that can quantitatively assess the feasibility and behaviour of the proposed models (SD) together, which should facilitate the increased impact of LSS deployments; this is the subject of on-going work by authors and sponsors.

7. Acknowledgements

The authors would like to thank the Economic and Social Research Council (UK) and Chris Rees and Mike Titchen of SigmaPro, for sponsoring and mentoring this research.

8. References

Ackoff, R.L. (2006) "Why Few Organizations Adopt Systems Thinking", *Systems Research and Behavioral Science Vol.* 23 *pp.* 705-708, ISSN 10927026.

Ackoff, R.L. (1995), "'Whole-ing' the Parts and Righting the Wrongs", *Systems Research. Vol.* 12 (1) *pp.* 43-46, ISSN 07317239

Ackoff, R.L. (1971), "Towards a System of Systems Concepts", *Management Science. Vol.*17 (11) *pp.* 661-671, ISSN 00251909

Agricola, G., "*De Re Metallica*". Froben, Basel. 1556. ISBN 1154680657

Akao, Y. Ed. *"Hoshin Kanri: Policy Deployment for Successful TQM"* Productivity Press, NY, 1991. ISBN 9781563273117

Anon, *Quality Magazine.*, BNP Media, 2010, *pp.*48-51

Balogun, J. & Johnson, G. (2005), "Intended Strategies to Unintended Outcomes", *Organization Studies. Vol.* 26 (11) *pp.*1573-1601, ISSN 01708406

Babbage, C. *"On the Economy of Machinery and Manufactures"*, Charles Knight, Pall Mall, London 1832. ISBN 0678000018

Baines, T. Lightfoot, H. Williams, G. M. & Greenough, R. (2006),"State-of-the-art in lean design engineering: a literature review on white collar lean", *Journal of Engineering Manufacture. Vol.* 220 (9) *pp.* 1539-1547, ISSN 09544054.

Battista, J.R., 'The Holistic Paradigm and General Systems Theory', *General Systems,* XXII, 1977.

Bertalanffy, L. von, *General Systems Theory.* Brazillier, New York. 1968, ISBN 0807604534

Beer, S. *Diagnosing the Systems for Organisations.* Wiley, Chichester. 1985.

Bendoly, E. & Hur, D. (2007), "Bipolarity in reactions to operational 'constraints': OM bugs under an OB lens.", *Journal of Operations Management. Vol.* 25 *pp.*1-13, ISSN 02726963

Boudreau, J. Hopp, W. McClain, J.O. & Thomas, L.J. (2003) "On the interface between operations and human resources management", *Manufacturing and Service Operations Management. Vol.* 5 (3) *pp.*179-202, ISSN 15234614

Boulding, K., (1956) 'General Systems Theory - the Skeleton of Science'. *Management Science.* Vol. 2. pp 197-208, ISSN 00251909

Brady. J. E. & Allen T.T. (2006), "Six Sigma Literature: A Review and Agenda for Future Research", *Quality and Reliability Engineering International. Vol.* 22 (3) *pp.*335-367, ISSN 07488017

Burgess, D., *Quality Magazine.* BNP Media, 2010, *pp.* 42-45

Chakravorty, S.S. (2009), "Six Sigma programs: An implementation model." *International Journal of Production Economics. Vol.* 119 (1) *pp.*1-16, ISSN 09255273.

Checkland, P.B., (1988) 'Soft Systems Methodology: An Overview'. *J. of Applied Systems Analysis.* Vol. 15. pp27-30, ISSN 03089541

Checkland, P. & Poulter, J. *Learning for action: a short definitive account of soft systems methodology and its use for practitioners, teachers and students.* Wiley. 2006, ISBN 0470025549

Checkland, P. & Scholes, J. "Soft Systems Methodology in Action", Wiley, 1990, ISBN 0471986054

Churchman, C.W., '*The Systems Approach*'. Dell Publishing. NY. pp11. 1968, ISBN 0440384079

Clegg BT, (2007) *'Building a Holarchy using Business Process Orientated Holonic (PrOH) Modeling'*. IEEE Systems, Man and Cybernetics: Part A. Vol.37, No.1. pp.23-40, ISSN 10834427

Clegg, B. Rees, C. & Titchen, M. (2010), "A study into the effectiveness of quality management training: A focus on tools and critical success factors", *The TQM Journal. Vol.* 22 (2) *pp.* 188-208, ISSN 17542731

Conti, T. (2010), "Systems thinking in quality management", *The TQM Journal. Vol.* 22 (4) *pp.* 352-368, ISSN 17542731

Forrester, J.W., *Industrial Dynamics*. Waltham: Pegasus Communication, 1961, ISBN 1883823366

Forrester, J.W. (1994), "System dynamics, system thinking and soft OR", *System Dynamics Review. Vol.* 10 (2-3) *pp.* 245-256, ISSN 08837066.

Flood, R.L. & Jackson, M.C. *Creative Problem Solving: Total Systems Intervention.* Wiley, Chichester, 1991, ISBN 0471930520

Ford, H., "*Today and Tomorrow*". Productivity Press, Portland Oregon, 2003

Furterer, S. & Elshennawy, A.K. (2005), "Implementation of TQM and Lean Six Sigma tools in local government: A framework and a case study", *Total Quality Management and Business Excellence. Vol.* 16 (10) *pp.*1179-1191.

George, M. L. "*Lean Six Sigma: Combining Six Sigma Quality with Lean Production Speed*", Blacklick, OH, USA: McGraw-Hill Professional Publishing, 2002

Houghton, L. (2009). "Generalization and Systemic Epistemology: Why should it Make Sense?", *Systems Research and Behavioral Science. Vol.* 26 *pp.* 99-108, ISSN 10927026.

Jackson, M.C. (1990). 'Beyond a System of Systems Methodologies'. *The Journal of the Operational Research Society. Vol.* 41 (8) *pp.* 657-668, ISSN 01605682

Jackson, M.C., (1995). 'Beyond the Fads; Systems Thinking for Managers'. *Systems Research.* Vol. 12. No. 1, pp25-42, ISSN 10991735.

Jayawarna, D. & Holt, R. (2009), "Knowledge and quality management: An R&D perspective", *Technovation. Vol.* 29 (11) *pp.* 775-785, ISSN 01664972

Kuei, C-H. & Madu, C.N. (2003), "Customer-centric six sigma quality and reliability management", *International Journal of Quality and Reliability Management. Vol.* 20 (8) *pp.* 954-964, ISSN 0265671X

Kiani, B., Shirouyehzad, H., Bafti, F.K., Fouladgar, H. (2009), "System dynamics approach to analysing the cost factors effects on cost of quality" *International Journal of Quality & Reliability Management*, Vol. 26 (7)

Kwak, Y.H. & Anbari, F.T. (2006), "Benefits, obstacles and future of six sigma approach", *Technovation. Vol.* 26 (5/6) *pp.* 708-715, ISSN 01664972

Lane, D.C. & Oliva, R. (1998), "The greater whole: Towards a synthesis of systems dynamics and soft systems methodology", *European Journal of Operational Research. Vol.* 107 *pp.* 214-235, ISSN 03772217.

Lane, D.C. (1999), "Social theory and systems dynamics practice", *European Journal of Operational Research. Vol.* 113 *pp.* 501-527, ISSN 03772217.

Ledington, P.W.J & Ledington, J. (1999). 'The Problem of Comparison in Soft Systems Methodology'. *Systems Research and Behavioral Science. Vol.* 16 *pp.* 329-339, ISSN 10927026.

Liker, J.K. "*The Toyota way: 14 management principles from the world's greatest manufacturer*" McGraw-Hill. 2004

Linderman, K. Schroeder, R.G. & Choo, A.S. (2006), "Six Sigma: The role of goals in improvement teams", *Journal of Operations Management. Vol.* 24 *pp.* 779-790, ISSN 02726963

Lu, D. & Betts, A. (2011), "Why process improvement training fails.", *Journal of Workplace Learning. Vol.* 23 (2) *pp.* 117-132, ISSN 13665626

Martin, J.W. "*Lean Six Sigma for Supply Chain Management: The 10 Step Solution Process*", McGraw-Hill Professional. 2006, ISBN 0071479422

McAdam. R. & Hallet. S.A. (2010), "An absorptive capacity interpretation of Six Sigma.", *Journal of Manufacturing Technology Management. Vol.* 21 (5) *pp.* 624-645, ISSN 1741038X

McAdam. R. & Lafferty. B. (2004), "A multilevel case study critique of Six Sigma: Statistical control or strategic change.", *International Journal of Operations and Production Management. Vol.* 24 (5/6) *pp.* 530-549, ISSN 01443577

McComb, S.A. Green, S.G. & Compton, W.D. (2007), "Team flexibility's relationship to staffing and performance in complex projects: An empirical analysis", *Journal of Engineering Technology Management. Vol.* 15 (5) *pp.* 293-313, ISSN 09234748

Mingers, J. (2011) "Soft OR comes of age - but not everywhere", *Omega. Vol.* 39 (6) *pp.* 729-741, ISSN 03050483.

Mingers, J. (2003) "A classification of the philosophical assumptions of management science methods", *Journal of the Operational Research Society. Vol.* 54 *pp.* 559-570, ISSN 01605682.

Mingers, J. & Rosenhead, J. (2004), "Problem Structuring Methods in Action", *European Journal of Operational Research. Vol.* 152 (3) *pp.* 530-554, ISSN 03772217.

Mingers, J. & Taylor, S. (1992), "The Use of Soft Systems Methodology in Practice", *Journal of the Operational Research Society. Vol.* 43 (4) *pp.* 321-332, ISSN 01605682.

Mintzberg, H. (1971) "Analysis from observation", *Management Science Vol.* 18 (3), pp. B97-B110 ISSN 00251909

Munro, I. & Mingers, J. (2002), "The use of multi-methodology in practice - results of a survey of practitioners", *Journal of the Operational Research Society. Vol.* 53 *pp.*369-378, ISSN 01605682

Naslund. N. (2008), "Lean, Six Sigma and Lean Sigma: Fads or real process improvement methods?", *Business Process management Journal. Vol.* 14 (3) *pp.* 269-287, ISSN 14637154

Naveh, E. (2007), "Formality and discretion in successful R&D projects", *Journal of Operations Management. Vol.* 25 *pp.* 110-125, ISSN 02726963

Nonthaleerak. P. & Hendry. L. (2008), "Exploring the Six Sigma phenomenon using multiple case study evidence", *International journal of Operations and Production Management. Vol.* 20 (3) *pp.* 279-303, ISSN 01443577

Pande, P.S., "*The Six Sigma Way: how GE, Motorola, and other top companies are honing their performance*", McGraw-Hill. 2000, ISBN 9780071358064

Peery, N.S. (1972) "General Systems Theory: An Enquiry into its Social Philosophy", *The Academy of Management Journal. Vol.* 15 (4) *pp.* 495-510, ISSN 19480989

Pavlov, A. & Bourne, M. (2011), "Explaining the effects of performance measurement on performance and organizational routines perspective", *International Journal of Operations and Production Management. Vol.* 31 (1) *pp.* 101-122, ISSN 01443577

Pepper, M.P.J and Spedding, T.A. (2010) 'The evolution of Lean Six Sigma', *International Journal of Quality & Reliability Management*, Vol. 27 (2), pp.138-155

Proudlove, N. Moxham, C. & Boaden, R. (2008), "Lessons for lean in healthcare from using Six Sigma in the NHS", *Public Money and Management. Vol.* 28 (1) *pp.*27-34 ISSN 09540962

Rice, R.E. (2008), "Unusual Routines: Organizational (Non)Sensemaking", *Journal of Communication. Vol.* 58 *pp.* 1-19. ISSN 00219916

Robbins, S.P. "*Organizational Behavior*", 9th Ed., Prentice-Hall Inc., New Jersey. 2001, ISBN 0130617210

Rodrigues, A. & Bowers, J. (1996) "Systems dynamics in project management: A comparative analysis with traditional methods", *System Dynamics Review. Vol.* 12 (2) *pp.* 121-139, ISSN 08837066

Schwaninger, M. (1997). 'Integrative Systems Methodology: Heuristic for Requisite Variety', *International Transactions in Operational Research. Vol.* 4 (4) *pp.* 109-123, ISSN 09696016.

Senge, P., '*The Fifth Discipline: The art and Practice of the learning Organisation*'. Doubleday/ Currency. NY. 1990, IBSN 0385517254

Shah, R. & Ward, P.T. (2007), "Defining and developing measures of lean production", *Journal of Operations Management. Vol.* 25 *pp.* 758-805, ISSN 02726963

Skinner, W. (2007) Manufacturing strategy: the story of its evolution, *Journal of Operations Management Vol.* 25 (2), pp. 328–335 ISSN 02726963

Sprague, L.G. (2007), "Evolution of the field of operations management", J*ournal of Operations Management. Vol.* 25 (2) *pp.*219-238, ISSN 02726963

Su. C. & Chou. C. (2008), "A systematic methodology for the creation of six sigma projects: a case of a semiconductor foundry", *Expert Systems with Applications. Vol.* 34 (4) *pp.* 2693-2703, ISSN 09574174

Taylor, F.W., "Shop Management", Harper & Brothers Publishers, New York and London 1911, ISBN 1153738376

Thomas, A. Barton, R. & Chuke-Okafor. C. (2009), "Applying Lean Six Sigma in a small engineering company – a model for change", *Journal of Manufacturing Technology Management. Vol.* 20 (1) *pp.*113-129. ISSN 1741038X

Tjahjono. B. Ball. P. Vitanov. V.I. Scorzafave. C. Noguiera. J. Calleja. J. Minguet. M. Narasimha. L. Rivas. A. Srivastava. A. Srivastava. S. & Yadav. A. (2010), "Six Sigma: a literature review", *International Journal of Lean Six Sigma. Vol.*1 (3) *pp.* 216-233, ISSN 20404166

Tushman, M. & O'Rielly, C. (2007). 'Research and Relevance: Implications of Pasteur's Quadrant for Doctorial Programs and Faculty Development', *Academy of Management Review. Vol.* 50 (4) *pp.* 769-774, ISSN 03637425

Waelchli, F. (1992) 'Eleven Theses of General Systems Theory (GST)'. *Systems Research*. Vol. 9, No. 4, pp3-8. ISSN 07317239

Weinburg, G.M., '*An Introduction to General Systems Thinking*'. Wiley, NY. 1975, ISBN 0932633498

Wiklund. H & Wiklund. P.S. (2002), "Widening the Six Sigma concept: An approach to improve organizational learning", *Total Quality Management & Business Excellence. Vol.*13 (2) *pp.* 233-239, ISSN 14783363

White, L. (2009), "Understanding problem structuring methods interventions", *European Journal of Operational Research." Vol.*199 *pp.* 823-833, ISSN 03772217.

Womack, J.P., Jones, D.T. & Roos, D., "*The Machine That Changed the World.*", 1st ed., 1990, Rawson Associates, Macmillan Publishing Associates, New York, 1990. ISBN 0892563508

Yang. T. & Hsieh. C. (2009), "Six Sigma project selection using national quality award criteria and Delphi fuzzy multiple criteria decision-making method", *Expert Systems with Applications. Vol.* 36 (4) *pp.* 7594-7603, ISSN 09574174

Zhang, H. (2010). 'Soft Systems Methodology and 'Soft' Philosophy of Science', *Systems Research and Behavioral Science Vol.27 pp.* 156-170, ISSN 10927026.

Zu, X. Robbins, T.L. & Fredendall, L.D. (2010), "Mapping the Critical Links Between Organizational Culture and TQM / Six Sigma Practices", *International Journal of Production Economics. Vol.* 123 (1) *pp.* 86-106, ISSN 09255273.

4

The Data Distribution Service – The Communication Middleware Fabric for Scalable and Extensible Systems-of-Systems

Angelo Corsaro[1] and Douglas C. Schmidt[2]
[1]*PrismTech*
[2]*Vanderbilt University*
USA

1. Introduction

During the past several decades techniques and technologies have emerged to design and implement distributed systems effectively. A remaining challenge, however, is devising techniques and technologies that will help design and implement SoSs. SoSs present some unique challenges when compared to traditional systems since their scale, heterogeneity, extensibility, and evolvability requirements are unprecedented compared to traditional systems Northrop et al. (2006).

For instance, in Systems-of-Systems (SoS), such as the one depicted in Figure 1, the computational and communication resources involved are highly heterogeneous, which yields situations where high-end systems connected to high-speed networks must cooperate with embedded devices or resource-constrained edge systems connected over bandwidth-limited links. Moreover, in SoS it is common to find multiple administrative entities that manage the different parts, so upgrading the system must be incremental and never require a full redeployment of the whole SoS. In addition, SoS are often characterized by high degrees of dynamism and thus must enable subsystems and devices dynamically joining and leaving the federation of system elements.

The Object Management Group (OMG) Data Distribution Service for Real-Time Systems (DDS) Group (2004) is a standard for data-centric Publish/Subscribe (P/S) introduced in 2004 to address the challenges faced by important mission-critical systems and systems-of-systems. As described in the reminder of this Chapter, DDS addresses all the key challenges posed by SoS outlined above. As a result, it provides the most natural choice as the communication middleware fabrice for developing scalable and extensible SoS.

Since its inception DDS has experienced a swift adoption in several different domains. The reason for this successful adoption stems largely from its following characteristics:

1. DDS has been designed to scale up and down, allowing deployments that range from resource-constrained embedded systems to large-scale systems-of-systems.
2. DDS is equipped with a powerful set of QoS policies that allow applications fine-grain control over key data distribution properties, such as data availability, timeliness, resource consumption, and usage.

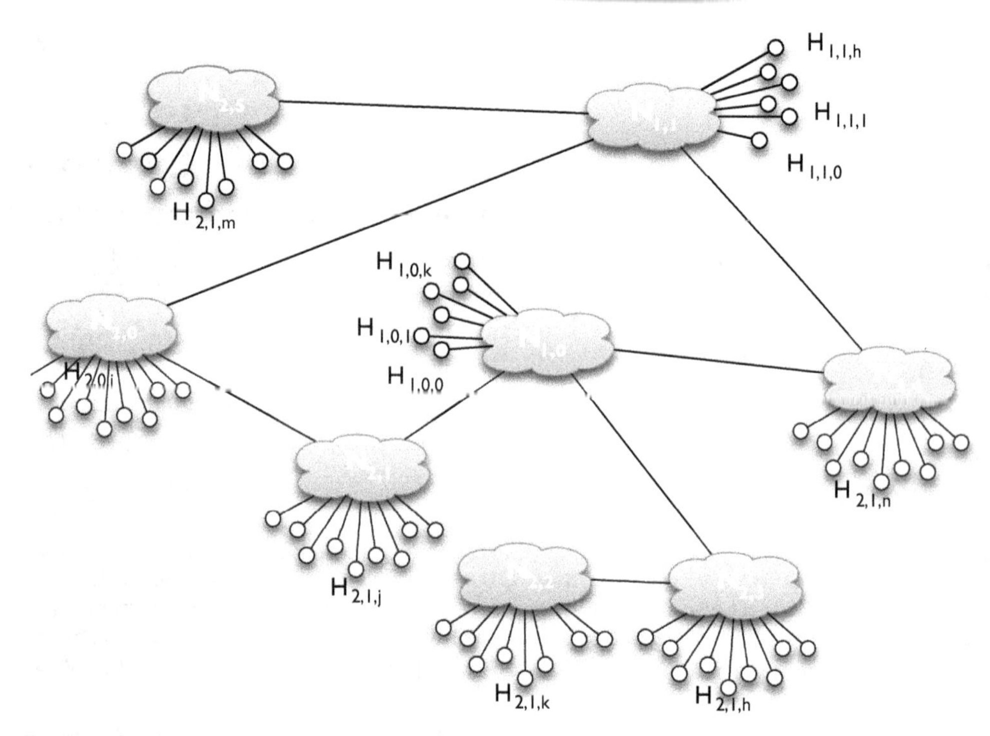

Fig. 1. A System-of-Systems Architecture.

3. DDS is equipped with a powerful type system that provides end-to-end type safety for built-in and user-defined types, as well as type-safe extensibility.

As a result of these characteristics, the OMG DDS standard is the most advanced data distribution technology and is a key building-block for many existing and upcoming SoSs.

The remainder of this chapter presents an in-depth introduction to DDS, as well a set of guidelines on how to apply this technology to architect scalable and efficient SoSs. The chapter concludes with a preview of forthcoming DDS innovations.

2. Overview of the OMG Data Distribution Service (DDS)

P/S is a paradigm for one-to-many communication that provides anonymous, decoupled, and asynchronous communication between producers of data–the publishers–and consumers of data–the subscribers. This paradigm is at the foundation of many technologies used today to develop and integrate distributed applications (such as social application, e-services, financial trading, etc.), while ensuring the composed parts remain loosely coupled and independently evolvable.

Different implementations of the P/S abstraction have emerged to support the needs of different application domains. DDS Group (2004) is an OMG P/S standard that enables scalable, real-time, dependable and high performance data exchanges between publishers and subscribers. DDS addresses the needs of mission- and business-critical applications, such as, financial trading, air traffic control and management, defense, aerospace, smart grids, and complex supervisory and telemetry systems. That key challenges addressed by DDS are to provide a P/S technology in which data exchanged between publishers and subscribers are:

- **Real-time**, meaning that the right information is delivered at the right place at the right time–all the time. Failing to deliver key information within the required deadlines can lead to life-, mission- or business-threatening situations. For instance in financial trading 1ms can make the difference between losing or gaining $1M. Likewise, in a supervisory applications for power-grids, failing to meet deadlines under an overload situation could lead to severe blackout, such as the one experienced by the northeastern US and Canada in 2003 *http://bit.ly/oKmbhM* (2003).
- **Dependable**, thus ensuring availability, reliability, safety and integrity in spite of hardware and software failures. For instance, the lives of thousands of air travelers depend on the reliable functioning of an air traffic control and management system. These systems must ensure 99.999% availability and ensure that critical data is delivered reliably, regardless of experienced failures.
- **High-performance**, which necessitates the ability to distribute very high volumes of data with very low latencies. As an example, financial auto-trading applications must handle millions of messages per second, each delivered reliably with minimal latency, e.g., on the order of tens of microseconds.

2.1 Components in the DDS standard

The components in the OMG DDS standards family are shown in Figure 2 and consist of the DDS v1.2 API Group (2004) and the Data Distribution Service Interoperability Wire Protocol (DDSI) Group (2006). The DDS API standard ensures source code portability across different vendor implementations, while the DDSI Standard ensures on the wire interoperability across DDS implementations from different vendors. The DDS API standard shown in Figure 2 also defines several profiles that enhance real-time P/S with content filtering, persistence, automatic fail-over, and transparent integration into object oriented languages.

The DDS standard was formally adopted by the OMG in 2004. It quickly became the established P/S technology for distributing high volumes of data dependably and with predictable low latencies in applications such as radar processors, flying and land drones, combat management systems, air traffic control and management, high performance telemetry, supervisory control and data acquisition systems, and automated stocks and options trading. Along with wide commercial adoption, the DDS standard has been mandated as the technology for real-time data distribution by organization worldwide, including the US Navy, the Department of Defence (DoD) Information-Technology Standards Registry (DISR) the UK Mistry of Defence (MoD), the Military Vehicle Association (MILVA), and EUROCAE–the European organization that regulates standards in Air Traffic Control and Management.

2.2 Key DDS architectural concepts and entities

Below we summarize the key architectural concepts and entities in DDS.

2.2.1 Global data space

The key abstraction at the foundation of DDS is a fully distributed Global Data Space (GDS) (see Figure 3). The DDS specification requires a fully distributed implementation of the GDS

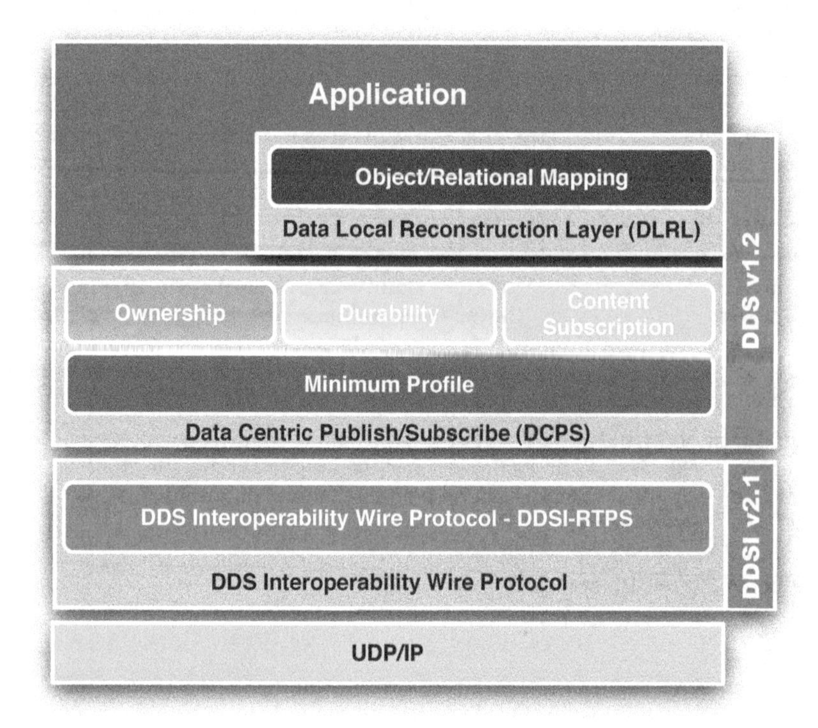

Fig. 2. The DDS Standard.

to avoid single points of failure or single points of contention. Publishers and Subscribers can join or leave the GDS at any point in time as they are dynamically discovered. The dynamic discovery of Publisher and Subscribers is performed by the GDS and does not rely on any kind of centralized registry, such as those found in other P/S technologies like the Java Message Service (JMS) Microsystems (2002). The GDS also discovers application defined data types and propagates them as part of the discovery process.

Since DDS provides a GDS equipped with dynamic discovery, there is no need for applications to configure anything explicitly when a system is deployed. Applications will be automatically discovered and data will begin to flow. Moreover, since the GDS is fully distributed the crash of one server will not induce unknown consequences on the system availability, i.e., in DDS there is no single point of failure and the system as a whole will continue to run even if applications crash/restart or connect/disconnect.

2.2.2 Topic

The information that populates the GDS is defined by means of DDS *Topics*. A *topic* is identified by a unique name, a data type, and a collection of Quality of Service (QoS) policies. The unique name provides a mean of uniquely referring to given topics, the data type defines

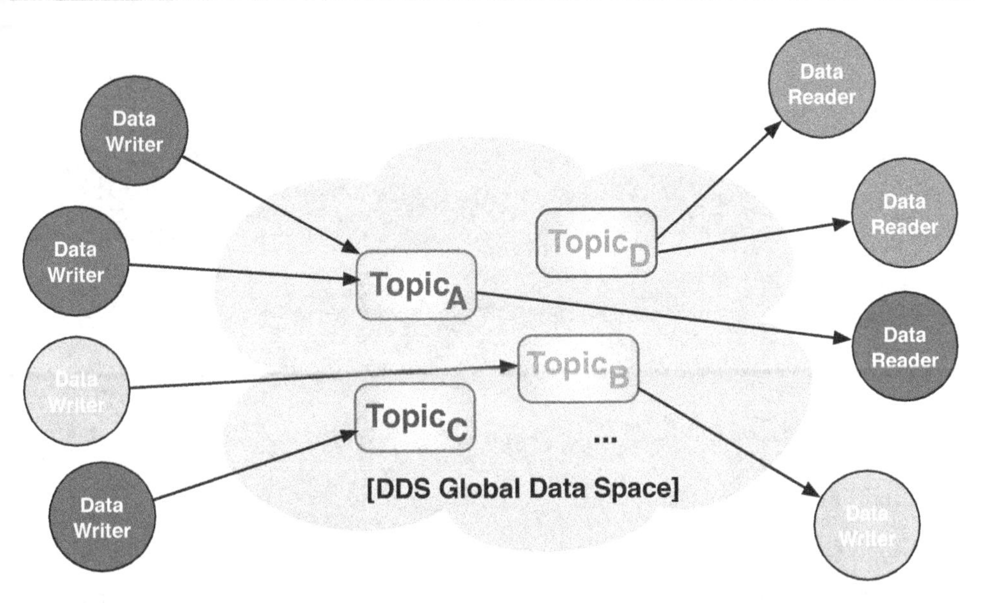

Fig. 3. The DDS Global Data Space.

the type of the stream of data, and the QoS captures all the non-functional aspect of the information, such as its temporal properties or its availability.

2.2.2.1 Topic types.

DDS Topics can be specified using several different syntaxes, such as Interface Definition Language (IDL), eXtensible Markup Langauge (XML), Unified Modeling Langauge (UML), and annotated Java. For instance, Listing 1 shows a type declaration for an hypothetical temperature sensor topic-type. Some of the attributes of a topic-type can be marked as representing the key of the type.

Listing 1. Topic type declaration for an hypothetical temperature sensor

```
1 struct TempSensorType {
    @Key
3   short id;
    float temp;
5   float hum;
  };
```

The key allows DDS to deal with specific instances of a given topic. For instance, the topic type declaration in Listing 1 defines the `id` attributes as being the key of the type. Each unique `id` value therefore identifies a specific topic instance for which DDS will manage the entire life-cycle, which allows an application to identify the specific source of data, such as the specific physical sensor whose `id=5`. Figure 4 provides a visual representation of the relationship existing between a topic, its instances, and the associated data streams.

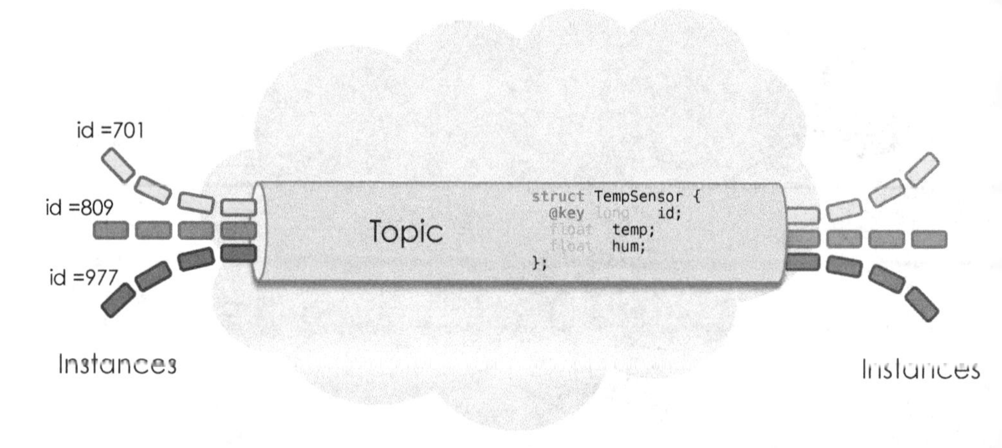

Fig. 4. Topic Instances and Data Streams.

2.2.2.2 Topic QoS.

The Topic QoS provides a mechanism to express relevant non-functional properties of a topic. Section 4 presents a detailed description of the DDS QoS model, but at this point we simply mention the ability of defining QoS for topics to capture the key non-functional invariant of the system and make them explicit and visible.

2.2.2.3 Content filters.

DDS supports defining content filters over a specific topic. These content filters are defined by instantiating a `ContentFilteredTopic` for an existing topic and providing a filter expression. The filter expression follows the same syntax of the WHERE clause on a SQL statement and can operate on any of the topic type attributes. For instance, a filter expression for a temperature sensor topic could be `"id = 101 AND (temp > 35 OR hum > 65)"`.

2.2.3 DataWriters and DataReaders

Since a topic defines the subjects produced and consumed, DDS provides two abstractions for writing and reading these topics: *DataWriters*s and *DataReader*s, respectively. Both DataReaders and DataWriters are strongly typed and are defined for a specific topic and topic type.

2.2.4 Publishers, Subscribers and Partitions

DDS also defines *Publishers* and *Subscribers*, which group together sets of DataReaders and DataWriters and perform some coordinated actions over them, as well as manage the communication session. DDS also supports the concept of a *Partition* which can be used to organize data flows within a GDS to decompose un-related sets of topics.

2.3 Example of applying DDS

Now that we presented the key architectural concepts and entities in DDS, we will show the anatomy of a simple DDS application. Listing 2 shows the steps required to join a DDS domain, to define a publisher and a topic on the given domain, and then create a data writer and publish a sample.

Listing 2. A simple application writing DDS samples.

```
// -- Joing Domain by creating a DomainParticipant
int32_t domainID = 0;
DomainParticipant dp(domainID);

// -- Get write acces to a domain by creating a Publisher
Publisher pub(dp);

// -- Register Topic
Topic<TempSensor> topic(dp, "TemSensorTopic");

// -- Create DataWriter
DataWriter<TempSensor> dw(pub,topic);

// -- Writer a sample
dw <<  TempSensor(701, 25, 67);
```

Note that no QoS has been specified for any of the DDS entities defined in this code fragment, so the behavior will be the default QoS.

Listing 3. A simple application reading DDS samples.

```
// -- Joing Domain by creating a DomainParticipant
int32_t domainID = 0;
DomainParticipant dp(domainID);

// -- Get write acces to a domain by creating a Subscriber
Subscriber sub(dp);

// -- Register Topic
Topic<TempSensor> topic(dp, "TemSensorTopic");

// -- Create DataReader
DataReader<TempSensor> dr(pub,topic);


std::vector<TempSensor> data(MAX_LEN);
std::vector<SampleInfo> info(MAX_LEN);
// -- Rear a samples
dr.read(data.begin(), info.begin(), MAX_LEN)
```

Listing 3 shows the steps required to read the data samples published on a given topic. This code fragment shows the application proactively reading data. DDS also provides a notification mechanism that informs a datareader via a callback when new data is available, as well as an operation for waiting for new data to become available.

3. The DDS type system

Strong typing plays a key role in developing software systems that are easier to extend, less expensive to maintain, and in which runtime errors are limited to those computational aspects that either cannot be decided at compile-time or that cannot be detected at compile-time by the type system. Since DDS was designed to target mission- and business-critical systems where safety, extensibility, and maintainability are critically important, DDS adopted a strong and statically typed system to define type properties of a topic. This section provides an overview of the DDS type system.

3.1 Structural polymorphic types system

DDS provides a *polymorphic structural type system* Cardelli (1996); Cardelli & Wegner (1985); Group (2010), which means that the type system not only supports polymorphism, but also bases its sub-typing on the *structure* of a type, as opposed than its *name*. For example, consider the types declared in Listing 4.

Listing 4. Nominal vs. Structural Subtypes

```
  struct Coord2D {
2   int x;
    int y;
4 };

6 struct Coord3D : Coord2D {
    int z;
8 };

10 struct Coord {
    int x;
12  int y;
    int z;
14 };
```

In a traditional nominal polymorphic type system, the `Coord3D` would be a subtype of `Coord2D`, which would be expressed by writing *Coord3D* <: *Coord2D*. In a nominal type system, however, there would be no relationship between the `Coord2D/Coord3D` with the type `Coord`. Conversely, in a polymorphic structural type system like DDS the type `Coord` is a subtype of the type `Coord2D`—thus *Coord* <: *Coord2D* and it is structurally the same type as the type `Coord3D`.

The main advantage of a polymorphic structural type system over nominal type systems is that the former considers the structure of the type as opposed to the name to determine sub-types relationships. As a result, polymorphic structural types systems are more flexible and well-suited for SoS. In particular, types in SoS often need to evolve incrementally to provide a new functionality to most subsystems and systems, *without* requiring a complete redeploy of the entire SoS.

In the example above the type `Coord` is a monotonic extension of the type `Coord2D` since it adds a new attribute "at the end." The DDS type system can handle both attribute reordering and attribute removal with equal alacrity.

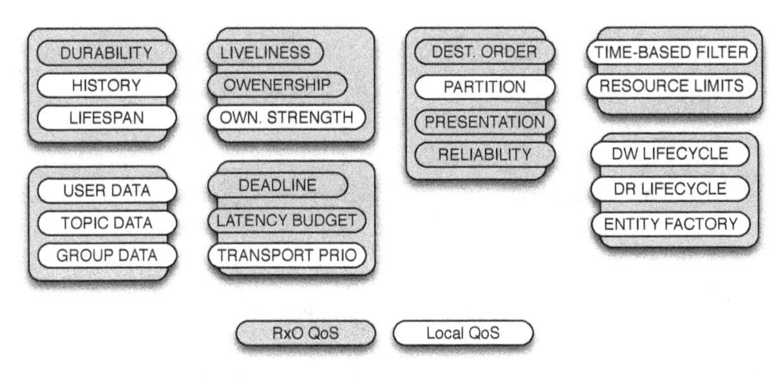

Fig. 5. DDS QoS Policies.

3.2 Type annotations

The DDS type systems supports an annotation system very similar to that available in Java Gosling et al. (2005). It also defines a set of built-in annotations that can be used to control the extensibility of a type, as well as the properties of the attributes of a given type. Some important built-in annotations are described below:

- The `@ID` annotation can be used to assign a global unique ID to the data members of a type. This ID is used to deal efficiently with attributes reordering.
- The `@Key` annotation can be used to identify the type members that constitute the key of the topic type.
- The `@Optional` annotations can be used to express that an attribute is optional and might not be set by the sender. DDS provides specific accessors for optional attributes that can be used to safely check whether the attribute is set or not. In addition, to save bandwidth, DDS will not send optional attributes for which a value has not been provided.
- The `@Shared` annotation can be used to specify that the attribute must be referenced through a pointer. This annotations helps avoid situations when large data-structures (such as images or large arrays) would be allocated contiguously with the topic type, which may be undesirable in resource-contrained embedded systems.
- The `@Extensibility` annotation can be used to control the level of extensibility allowed for a given topic type. The possible values for this annotation are (1) `Final` to express the fact that the type is sealed and cannot be evolved – as a result this type cannot be substituted by any other type that is structurally its subtype, (2) `Extensible` to express that only monotonic extensibility should be considered for a type, and (3) `mutable` to express that the most generic structural subtype rules for a type should be applied when considering subtypes.

In summary, the DDS type system provides all the advantages of a strongly-typed system, together with the flexibility of structural type systems. This combinations supports key requirements of a SoS since it preserves types end-to-end, while providing type-safe extensibility and incremental system evolution.

4. The DDS QoS model

DDS provides applications with explicit control over a wide set of non-functional properties, such as data availability, data delivery, data timeliness and resource usage through a rich set

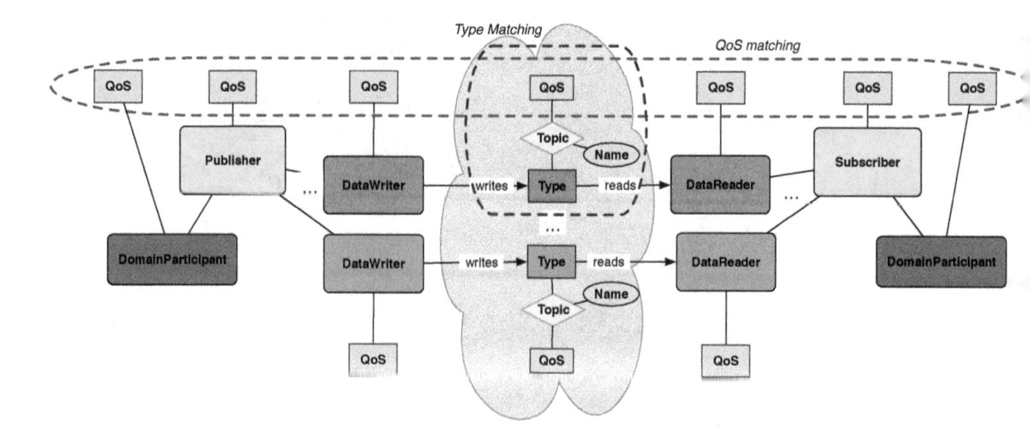

Fig. 6. DDS Request vs. Offered QoS Model.

of QoS policies – Figure 5 shows the full list of available QoS. The control provided by these policies over the key non-functional properties of data is important for traditional systems and indispensable for SoS. Each DDS entity (such as a topic, data reader, and data writer) can apply a subset of available QoS policies. The policies that control and end-to-end property are considered as part of the subscription matching. DDS uses a request vs. offered QoS matching approach, as shown in Figure 6 in which a data reader matches a data writer if and only if the QoS it is requesting for the given topic does not exceed (e.g., is no more stringent) than the QoS with which the data is produced by the data writer.

DDS subscriptions are matched against the topic type and name, as well as against the QoS being offered/requested by data writers and readers. This DDS matching mechanism ensures that (1) types are preserved end-to-end due to the topic type matching and (2) end-to-end QoS invariants are also preserved.

The reminder of this section describes the most important QoS policies in DDS.

4.1 Data availability

DDS provides the following QoS policies that control the availability of data to domain participants:

- The `DURABILITY` QoS policy controls the lifetime of the data written to the global data space in a DDS domain. Supported durability levels include (1) `VOLATILE`, which specifies that once data is published it is not maintained by DDS for delivery to late joining applications, (2) `TRANSIENT_LOCAL`, which specifies that publishers store data locally so that late joining subscribers get the last published item if a publisher is still alive, (3) `TRANSIENT`, which ensures that the GDS maintains the information outside the local scope of any publishers for use by late joining subscribers, and (4) `PERSISTENT`, which ensures that the GDS stores the information persistently so to make it available to late joiners even after the shutdown and restart of the whole system. Durability is achieved by relying on a durability service whose properties are configured by means of the `DURABILITY_SERVICE` QoS of non-volatile topics.

- The LIFESPAN QoS policy controls the interval of time during which a data sample is valid. The default value is infinite, with alternative values being the time-span for which the data can be considered valid.
- The HISTORY QoS policy controls the number of data samples (i.e., subsequent writes of the same topic) that must be stored for readers or writers. Possible values are the last sample, the last *n* samples, or all samples.

These DDS data availability QoS policies decouple applications in time and space. They also enable these applications to cooperate in highly dynamic environments characterized by continuous joining and leaving of publisher/subscribers. Such properties are particularly relevant in SoS since they increase the decoupling of the component parts.

4.2 Data delivery

DDS provides the following QoS policies that control how data is delivered and how publishers can claim exclusive rights on data updates:

- The PRESENTATION QoS policy gives control on how changes to the information model are presented to subscribers. This QoS gives control on the ordering as well as the coherency of data updates. The scope at which it is applied is defined by the access scope, which can be one of INSTANCE, TOPIC, or GROUP level.
- The RELIABILITY QoS policy controls the level of reliability associated with data diffusion. Possible choices are RELIABLE and BEST_EFFORT distribution.
- The PARTITION QoS policy gives control over the association between DDS partitions (represented by a string name) and a specific instance of a publisher/subscriber. This association provides DDS implementations with an abstraction that allow to segregate traffic generated by different partitions, thereby improving overall system scalability and performance.
- The DESTINATION_ORDER QoS policy controls the order of changes made by publishers to some instance of a given topic. DDS allows the ordering of different changes according to source or destination timestamps.
- The OWNERSHIP QoS policy controls whether it is allowed for multiple data writers to concurrently update a given topic instance. When set to EXCLUSIVE, this policy ensures that only one among active data writers–namely the one with the highest OWNERSHIP_STRENGTH–will change the value of a topic instance. The other writers, those with lower OWNERSHIP_STRENGTH, are still able to write, yet their updates will not have an impact visible on the distributed system. In case of failure of the highest strength data writer, DDS automatically switches to the next among the remaining data writers.

These DDS data delivery QoS policies control the reliability and availability of data, thereby allowing the delivery of the right data to the right place at the right time. More elaborate ways of selecting the right data are offered by the DDS content-awareness profile, which allows applications to select information of interest based upon their content. These QoS policies are particularly useful in SoS since they can be used to finely tune how—and to whom—data is delivered, thus limiting not only the amount of resources used, but also minimizing the level of interference by independent data streams.

4.3 Data timeliness

DDS provides the following QoS policies to control the timeliness properties of distributed data:

- The DEADLINE QoS policy allows applications to define the maximum inter-arrival time for data. DDS can be configured to automatically notify applications when deadlines are missed.
- The LATENCY_BUDGET QoS policy provides a means for applications to inform DDS how long time the middleware can take in order to make data available to subscribers. When set to zero, DDS sends the data right away, otherwise it uses the specified interval to exploit temporal locality and batch data into bigger messages so to optimize bandwidth, CPU and battery usage.
- The TRANSPORT_PRIORITY QoS policy allows applications to control the priority associated with the data flowing on the network. This priority is used by DDS to prioritize more important data relative to less important data.

  The DEADLINE, LATENCY_BUDGET and TRANSPORT_PRIORITY QoS policy provide the controls necessary to build priority pre-emptive distributed real-time systems. In these systems, the TRANSPORT_PRIORITY is derived from a static priority scheduling analysis, such as Rate Monotonic Analysis, the DEADLINE QoS policy represents the natural deadline of information and is used by DDS to notify violations, finally the LATENCY_BUDGET is used to optimize the resource utilization in the system.

These DDS data timeliness QoS policies provide control over the temporal properties of data. Such properties are particularly relevant in SoS since they can be used to define and control the temporal aspects of various subsystem data exchanges, while ensuring that bandwidth is exploited optimally.

4.4 Resources

DDS defines the following QoS policies to control the network and computing resources that are essential to meet data dissemination requirements:

- The TIME_BASED_FILTER QoS policy allows applications to specify the minimum inter-arrival time between data samples, thereby expressing their capability to consume information at a maximum rate. Samples that are produced at a faster pace are not delivered. This policy helps a DDS implementation optimize network bandwidth, memory, and processing power for subscribers that are connected over limited bandwidth networks or which have limited computing capabilities.
- The RESOURCE_LIMITS QoS policy allows applications to control the maximum available storage to hold topic instances and related number of historical samples DDS's QoS policies support the various elements and operating scenarios that constitute net-centric mission-critical information management. By controlling these QoS policies it is possible to scale DDS from low-end embedded systems connected with narrow and noisy radio links, to high-end servers connected to high-speed fiber-optic networks.

These DDS resource QoS policies provide control over the local and end-to-end resources, such as memory and network bandwidth. Such properties are particularly relevant in SoS since they are characterized by largely heterogeneous subsystems, devices, and network connections that often require down-sampling, as well as overall controlled limit on the amount of resources used.

4.5 Configuration

The QoS policies described above, provide control over the most important aspects of data delivery, availability, timeliness, and resource usage. DDS also supports the definition and distribution of user specified bootstrapping information via the following QoS policies:

- The `USER_DATA` QoS policy allows applications to associate a sequence of octets to domain participant, data readers and data writers. This data is then distributed by means of a built-in topic—which are topics pre-defined by the DDS standard and used for internal purposes. This QoS policy is commonly used to distribute security credentials.
- The `TOPIC_DATA` QoS policy allows applications to associate a sequence of octet with a topic. This bootstrapping information is distributed by means of a built-in topic. A common use of this QoS policy is to extend topics with additional information, or meta-information, such as IDL type-codes or XML schemas.
- The `GROUP_DATA` QoS policy allows applications to associate a sequence of octets with publishers and subscribers–this bootstrapping information is distributed by means built-in topics. A typical use of this information is to allow additional application control over subscriptions matching.

These DDS configuration QoS policies provide useful a mechanism for bootstrapping and configuring applications that run in SoS. This mechanism is particularly relevant in SoS since it provides a fully distributed means of providing configuration information.

5. Guidelines for building system of systems with DDS

This section presents a systematic method for building scalable, extensible, and efficient SoS by integrating them through DDS. This method has been used successfully in many SoS and has shown over time its value on addressing the key requirements faced when architecting a SoS, including (1) achieving the right level of scalability and extensibility while maintaining loose coupling and (2) minimizing resource usage.

The method described below will be presented as a series of steps, which are applied incrementally—and often iteratively–to achieve the appropriate SoS design. With time and experience the number of iterations required will diminish. It is advisable, however, to iterate two to three times through the steps described below when first applying the method. A visual representation of the steps involved in this approach is outlined in Figure 7.

5.1 Step I: Define the common information model

Integrating systems to form a SoS involves interconnecting these systems in a meaningful way. A conventional approach is to integrate systems in a pairwise manner, thus leading to a star-like system topology depicted in Figure 8(a). This conventional approach to system integration has several shortcomings, however, because (1) is not scalable since it requires integrating with $n-1$ systems, (2) it is not resource efficient since it duplicates information, and (3) it is hard to extend since each change is usually propagated on the $n-1$ point-to-point integrations.

An alternative approach to architect SoS involves focusing on a common set of abstractions—the *common information model*—used to represent all information that is necessary and relevant for the interworkings of the SoS. This approach, depicted in Figure 8(b), reduces—and in some cases eliminates—the integration effort since all systems

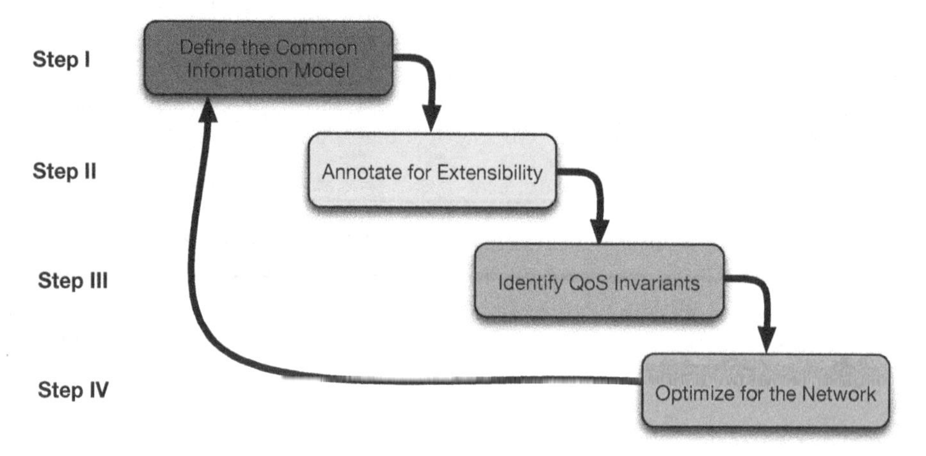

Fig. 7. Visual representation of the steps involved in the common information model approach.

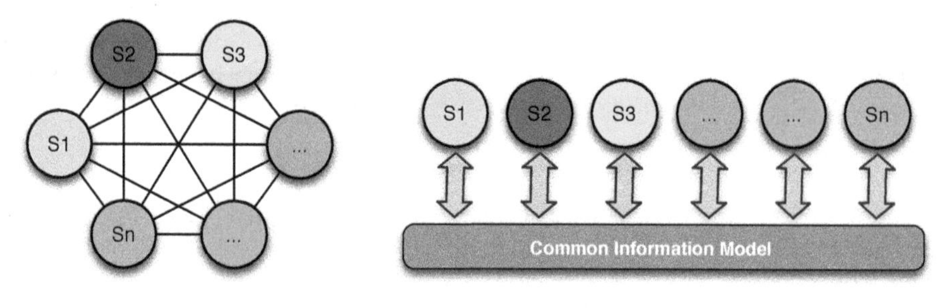

Fig. 8. System of System Integration Styles.

communicate using the same protocol and type system. This approach also migrates some effort to the careful design of the common information model, which defines the data representations that establish the *lingua franca* for the SoS.

The first step required to build a common information model involves devising the data types that capture the state and the events relevant throughout the SoS. This data can then be normalized using one of the several forms defined in the database management systems literature Ramakrishnan & Gehrke (2002). After applying this first step, the information model should be free of common anomalies, such as the update or deletion anomalies, and should be efficient, in the sense that information duplication will have been removed via the normalization process. At this point, however, the information model may not be ideal for SoS with respect to aspects like evolvability, efficiency, and QoS.

5.2 Step II: Annotate for extensibility

The second step of information model design should account for the extensibility requirements of each data type. Some data types must support evolutions, such as the ability of adding or removing attributes. Other data types must disallow extensions since they represent structural invariants, such as the type representing some physical structure of the system like the position and number of wheels. In either case, it is necessary to consciously decide what kind of extensibility is associated with each data type and use the annotations described in Section refSection:DDS:TypeSystem.

5.3 Step III: Identify QoS invariants

The previous steps allow refinement of an information model so that it cleanly captures key traits of the SoS. At this point, however, the information model does not capture any non-functional requirements associated with various data types. The next step, therefore, involves identifying the least stringent set of QoS policies (see Section 4) that should be associated with each data type to meet SoS non-functional requirements.

Decorating the common information model with the proper QoS ensures data producers can only produce data with stronger guarantees, whereas data consumers can only ask to consume data with weaker guarantees. This rule ensures the QoS violations do not occur and that the SoS will work as expected.

5.4 Step IV: Optimize for the network

After the common information model is decorated with QoS there is yet another steps to perform to address the fact that (1) a SoS is a distributed system, which requires awareness of network characteristics, and optimization of the used bandwidth as some of the subsystem or devices will often be connected through narrow-bandwidth links or will inherently have scarce computational and storage resources, and (2) DDS data is sent atomically, i.e., regardless of what changes occur in the (non-optional) fields of a data type when the entire data type is transmitted across the network.

These considerations requires additional scrutiny on the information model. In particular, it is necessary to identify all the data types that belong in one of the following cases:

- **Update frequency mix.** Each data type should be regarded with respect to the relative update frequency of its attributes. If there is a subset of attributes that are relatively static and another subset that changes relatively often, it is advisable to split the data type into two data types. The two types will share the key to allow correlation on the receiving side. This technique minimizes bandwidth utilization by limiting the amount of data sent over the network. For SoS that communicate over some low bandwidth links this technique significantly improve performance.
- **QoS mix.** Since QoS policies in DDS apply to the whole topic it is important to recognize that the `DURABILITY` or `RELIABILITY` QoS policy affects all attributes of the data type associated with the topic. In certain cases, however, some attributes will work fine with a weaker QoS setting. In such case, it is advisable to split data types into as many types as necessary to ensure that all attributes within a given data type share the same QoS, i.e., no attribute could select a weaker QoS without compromising correctness. This technique can improve both performance and resource utilization.

5.5 Step V: Iterate

The steps I to IV described above should be performed iteratively to ensure that (1) all key SoS concepts have been captured, (2) extensibility constraints have been handled, (3) the QoS policies properly capture non-functional invariants, and (4) the data model is efficient and scalable. With experience the number of iterations required will reduce, but you should typically apply these steps at least twice.

6. DDS: The road ahead

The DDS technology ecosystem is characterized by a lively and vibrant community that continues to innovate and extend the applicability of this powerful P/S technology. This section we summarize the state-of-the-art in DDS and then examine the DDS standards that will be forthcoming in the next year or so.

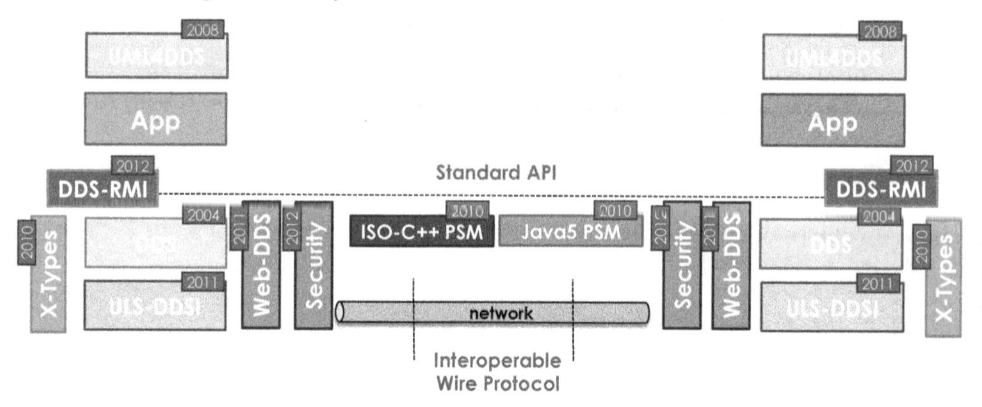

Fig. 9. The DDS Standard Evolution.

6.1 State-of-the-art for DDS

Figure 9 presents the whole family of standards belonging to the DDS ecosystem, including some of the upcoming extensions. As described in earlier sections, DDS supports QoS-enabled topic-based P/S Group (2004) where topics are strongly typed and where a structural polymorphic type system is used to enable type-safe system extensibility and evolution Group (2010). DDS is equipped with a standard wire-protocol, DDSI Group (2006), that provides native interoperability across vendor implementations of the standard.

Two new APIs were recently specified for DDS. One API defines a new mapping to ISO C++ and another defines a mapping to Java 5. Both APIs improve the productivity, safety and efficiency of DDS applications. As a result of these enhancements, DDS now provides the most dependable and productive standard-based communication middleware infrastructure for building mission- and business-critical SoS that require scalability, fault-tolerance, efficiency and performance.

6.2 Coming next

There are three areas that will yield the following new DDS standards by 2012:

- **Web-enabled DDS.** This specification addresses the needs of SoS that must expose DDS data to Web/Internet applications. These capabilities will make it possible to expose DDS topics in a standardized manner to both RESTful and W3C web services. Web-enabled DDS will simplify the way in which SoS can bring mission-critical and real-time information to enterprise applications, as well as to browser-enabled devices, such as smart phones and tablets.
- **Secure DDS.** To date, DDS Security has been an area where each vendor has provided their own proprietary (and thus non-interoperable) solutions. With the increased adoption of

DDS in SoS the need for an interoperable and extensible security infrastructure has become evident. As a result, work is progressing on an interoperable DDS security specification that will address many aspects of security, such as data confidentiality, integrity, assurance and availability. This specification is based on pluggable security modules that will allow vendors to provide a default interoperable set of behaviours and customers or vertical domains to develop their own customized security plugins.

- **Ultra Large Scale Systems (ULS) DDSI.** The DDS wire-protocol, known as DDSI, was designed by first optimizing for LAN deployments and then adding a series of mechanisms that vendors can use over WANs. The DDSI wire-protocol does not, however, take advantage of the latest research on dissemination and discovery protocols, such as encoding techniques, dynamic distribution trees, and distributed hash-tables. The ULS DDSI specification will thus extend the existing DDSI protocol to further improve its efficiency over WAN and improve the scalability on ULS deployments Northrop et al. (2006).

In summary, the DDS technology ecosystem continues to expand its applicability and support systems and SoS efficiently and effectively.

7. Concluding remarks

This chapter has introduced the DDS standard and explained its core concepts in the context of meeting the requirements of SoS. As it has emerged from the use cases cited throughout the chapter—as well as from the set of features characterizing this technology—DDS is the ideal technology for integrating Systems-of-Systems. The main properties DDS-based SoS enjoy include:

- **Interoperability and portability.** DDS provides a standardized API and a standardized wire-protocol, thereby enabling portability and interoperability of applications across DDS implementations. These capabilities are essential for SoS since it is hard to mandate a single product be used for all systems and subsystems, but it is easier to mandate a standard.
- **Loose coupling.** DDS completely decouples publishers and subscribers in both *time*, e.g., data readers can receive data that was produced before they had joined the system, and *space*, e.g., through its dynamic discovery that requires no specific configuration—applications dynamically discover the data and topics of interest. These two properties minimize coupling between the constituent parts of SoS, thereby enabling them to scale up and down seamlessly.
- **Extensibility.** The DDS type system provides built-in support for system extensibility and evolution. Moreover, the system information model can be evolved dynamically in a type-safe manner, which helps ensure key quality assurance properties in SoS.
- **Scalability, efficiency, and timeliness.** The DDS architecture and the protocols used in its core where designed to ensure scalability, efficiency, and performance. In addition, the QoS policies available in DDS provide fine-grained control over the non-functional properties of a system, thereby allowing finely tuning and optimization of its scalability, efficiency, and timeliness.

In summary, DDS is a natural choice as the integration infrastructure for SoS, as evidenced by the many adoptions of DDS as the basis for current and next-generation SoS.

8. Acronyms

DDS Data Distribution Service for Real-Time Systems 1

DDSI Data Distribution Service Interoperability Wire Protocol 3

DISR DoD Information-Technology Standards Registry 3

DoD Department of Defence 3

GDS Global Data Space 3

IDL Interface Definition Language 5

JMS the Java Message Service 4

MILVA Military Vehicle Association 3

MoD Mistry of Defence 3

OMG Object Management Group 1

P/S Publish/Subscribe 1

QoS Quality of Service 4

SoS Systems-of-Systems 1

ULS Ultra Large Scale Systems 17

UML Unified Modeling Langauge 5

XML eXtensible Markup Langauge 5

9. References

Cardelli, L. (1996). Type systems, *ACM Comput. Surv.* 28: 263–264.

Cardelli, L. & Wegner, P. (1985). On understanding types, data abstraction, and polymorphism, *ACM Computing Surveys* 17(4): 471–522.

Gosling, J., Joy, B., Steele, G. & Bracha, G. (2005). *Java(TM) Language Specification, The (3rd Edition) (Java (Addison-Wesley))*, Addison-Wesley Professional.

Group (2004). Data distribution service for real-time systems.

Group (2006). Data distribution service interoperability wire protocol.

Group (2010). Dynamic and extensible topic types.

http://bit.ly/oKmbhM (2003).

Microsystems, S. (2002). The java message service specification v1.1.

Northrop, L., Feiler, P., Gabriel, R. P., Goodenough, J., Linger, R., Longstaff, T., Kazman, R., Klein, M., Schmidt, D., Sullivan, K. & Wallnau, K. (2006). Ultra-Large-Scale Systems - The Software Challenge of the Future, *Technical report*, Software Engineering Institute, Carnegie Mellon.
URL: *http://www.sei.cmu.edu/uls/downloads.html*

Ramakrishnan, R. & Gehrke, J. (2002). *Database Management Systems*, 3rd edn, McGraw Hill Higher Education.

5

New Methods for Analysis of Systems-of-Systems and Policy: The Power of Systems Theory, Crowd Sourcing and Data Management

Alfredas Chmieliauskas, Emile J. L. Chappin, Chris B. Davis, Igor Nikolic and Gerard P. J. Dijkema
Delft University of Technology
The Netherlands

1. Introduction

Our world is a complex socio-technical system-of-systems (Chappin & Dijkema, 2007; Nikolic, 2009). Embedded within the geological, chemical and biological planetary context, the physical infrastructures, such as power grids or transport networks span the globe with energy and material flows. Social networks in the form of global commerce and the Internet blanket the planet in information flows. While parts of these global social and technical systems have been consciously engineered and managed, the overall system-of-systems (SoS) is emergent: it has no central coordinator or manager. The emergence of this socio-technical SoS has not been without consequences: the human species is currently facing a series of global challenges, such as resource depletion, environmental pollution and climate change. Tackling these issues requires active policy and management of those socio-technical SoS. But how are we to design policies if policy makers and managers have a limited span of control over small parts of the global system of systems?

The aim of this chapter is to discuss the roles novel applications of information technology and agent-based modeling have in understanding our world as a complex system and in designing effective policy. In other words, by bringing together systems theory and recent IT developments we can arrive at a better understanding of, and improved decision making on the operation and development of our complex socio-technical systems. We introduce a method for data-driven agent-based modeling, thus illustrate the power of the combination of systems theory and novel methods in data management, and present and discuss two case studies.

System-of-systems engineering (SoSe) methodologies only recently have been applied to policy analysis and design, even though policy analysis has its roots in systems thinking (Meadows, 2008). In this chapter, we continue on that path by further exploring a new combination of methods: serious games (Duke, 1980; Meadows, 1999), collaborative information management and agent-based modeling (Chappin, 2011; Epstein & Axtell, 1996; Nikolic, 2009).

We argue that this combination is especially useful for developing better understanding of our social and physical environment and its interaction with policy design and analysis. New technologies and methods in data management allow us to build better simulations by making use of real-world data. We create and use novel IT systems to bring together and mine relevant data. Patterns found in such observations are then used to create system abstractions – i.e. simplified and systematic perspectives or views on the system. The system abstractions are then incorporated into agent-based models that explore the dynamics in operation and long-term development. The socio-technical approach adds the social and institutional patterns to the technical view of the system examined. When trying to understand our infrastructures and industrial systems, human aspects cannot be ignored, nor be analyzed separately. The agent-based models allow us to explore the effects of policies on a highly interconnected system of systems.

A hybrid approach results – data plus models. We use this approach to explore the new gas market balancing regime in the Netherlands. A second case study deals with a long-term development of the Dutch power market (extending the work of Chappin, 2011; Chappin & Dijkema, 2010; Chappin et al., 2010). For both case studies models were developed using a new agent-based modeling platform *AgentSpring*[1]. The main feature of AgentSpring is its ability to handle large amounts of data. Models use a large number of data points as assumptions and input; models also produce a complex simulated world that consists out of hundreds of actors, thousands of things and millions of relations between them. AgentSpring incorporates new technologies that help mine the simulated world for patterns in the evolving system behavior.

We conclude by arguing that the new combination adds expression to the modeling effort, allowing to encode the complexity inherent in the system. More importantly, we hope to make models more transparent, duplicable, communicable, and understandable to the decision maker. The more comprehensive understanding of the relationships and patterns in the system modeled is one of the ways to increase the maturity of policy analysis and design.

2. Changing nature and application of SoSe research: more social

Systems thinking and system dynamics ideas were first applied in military research, enabled by breakthroughs in computing and fueled by cold war concerns (MIT, 1953). Similarly, traditional SoSe research has its roots in defense related areas. Three out of six SoSe definitions (Jamshidi, 2005) have their origins in military research. While the systems dynamics deals with the relations between the objects within the system (Forrester, 1961), SoSe goes further to understand and manage the relations between such systems. The primary goals of SoSe research have traditionally been overall SoS optimization, ensuring interoperability, reliability and minimizing human error in their functioning (Pei, 2000).

Just as the pioneer of system dynamics Jay Forrester went on to apply his military systems research insights in economics (Forrester, 1968), the SoS research is being transfered to and applied in social sciences. The SoSe practices and ideas are being extended into domains that deal with understanding and managing complex infrastructure: space exploration, transportation, energy and economics (DeLaurentis & Ayyalasomayajula, 2009). While these

[1] AgentSpring is open-source: `https://github.com/alfredas/AgentSpring`

systems are in many ways similar to defense systems, they have a strong social component that has to be taken into account. Arguably, the introduction of social aspects changes the fundamental qualities of the system and the research methods. When dealing with SoS that have explicit social elements we are faced with:

- deep uncertainty and unpredictability of
  - system boundaries and structure
  - system behavior and dynamics
- non-rational and multi-criteria optimization in decision making
- multiple stakeholders with multiple perspectives and values
- no single point of control

Even the very notion of an optimum becomes problematic – it depends on the stakeholder's point of view and the system boundaries that are porous, prone to change. The role of SoSe of socio-technical systems (STS) needs to change from optimizing system performance to assisting human decision making by identifying relevant data, simulating and replicating SoS operation and studying common patterns.

Such transition also means that the user of SoSe research is the decision maker, either of a large enterprise or, commonly – a policy maker. In the context of the following chapters the SoSe are discussed in the context of supporting policy making and terms SoS and STS are used interchangeably.

3. Policy and SoSe

Global infrastructure systems have always been complex, but it seems that only recently we are becoming aware of the true implications of this when designing policy. A semantic analysis of Journal of Systems Science and Systems Engineering (published in 2003-2010), for example, reveals an interesting trend. The term "policy" has been mentioned only in one abstract of the 33 articles published in 2003 (3%), while the same term has been used in 11 out of 26 articles published in 2010 (42%). An analysis based on a Google News search (from 2000-2010)[2] reveals a relation emerges between the terms "policy" and "system" within the mass media since 2008.

In the context of policy, we conjecture that SoS resarch is about the tools that allow us to grasp the complexity and experience the extent of our policy decisions. Such effects are evident in the complex relations between the 2011 earthquake in Japan and European energy flows in the future (i.e. as a consequence of the nuclear phase out in Germany), or the relations with iPad shipments in California (Los Angeles Times, 2011). As long as our understanding has changed we are conditioned to make decisions in a different way. This change can be illustrated by two stories about the same piece of infrastructure – the Moscow-Saint Petersburg railway.

The Moscow-Saint Petersburg railway, opened in 1851, follows a straight line except for a 17 km bend near the city of Novgorod. The widely believed urban myth states that the bend is a planning artifact. Tsar Nikolaev intended to draw a straight line joining the two cities, but had

[2] The methodology used in the analysis of Systems Science and Systems Engineering and Google News are described elsewhere (Kasmire et al., Submitted)

to draw around the finger holding the ruler to the map. Since then the bend has been know as the "tsar finger", even thought the real reasons for the bend were technical (Wikipedia, 2011a).

The second part to the story deals with the expansion of the railway after 150 years. The planned high speed track was canceled due to environmental protests over the fragile environment of the Valdai Hills.

The two stories (even though the first one is a myth, but a myth widely believed)(Wikipedia, 2011a) exemplify the differences between decision-making. While the Tsar could exercise absolute power and bend the railway, today the complex socio-technical constellation in many a country often requires much more subtle and informed decision making. The role of the decision maker is not only to optimize the technical performance (straight line – optimal) but also to reconcile the different interests and stakes, including the Valdai hills ecosystem preservation.

The role of SoSe is then to provide the methodologies and the tools to support such demanding decision-making process.

4. Proactive study of SoSe

DeLaurentis & Sindiy (2006) recommend a three-phase SoSe approach where a SoS problem is defined, abstracted and simulated. The definition phase consists of identifying and characterizing the SoS problem as it currently exists, seeing the problem in its context. The abstraction phase allows to identify actors, things and relations and gives inputs for the implementation phase. The implementation phase is meant for replicating and simulating the workings of the SoS. This framework is consistent with the role of SoSe for decision support. It also provides a useful skeleton to discuss new tools that are available to the SoSe researcher. Continuous development of these tools can be summarized as proactive study of SoSe.

Proactive study of SoSe focuses on continuous engineering of technical and social components of an information system that maps the complexity of the real-world SoSe. The primary focus of proactive study of SoSe are the new information technologies that can take the SoSe research further.

Proactive study of SoS involves creating an ecosystem of information management tools at researcher's disposal and identifying patterns when to use them and how to combine them to address a research problem. The following sections will identify relevant developments in information technologies, propose a set of useful tools and exemplify their application in the context of SoSe research. The nature of the tools discussed is highly technical but in the scope of this chapter it is the enabling effect of these tools that are important, not the technical implementation. We identify the growing abundance of data and the emergence of the Internet of Things, then we discuss the methods and tools how to use it in the context of SoSe analysis and in combination with models, simulations and games.

4.1 Big data

When Jay Forrester and his colleagues were tasked with creating a military information system half a century ago, one of the main challenges was to manage massive amount of information collected by the various radars (Everett et al., 1957). Later scientists researching ecosystems collected vast amounts of data expecting to mine it for patterns that would allow them to understand and simulate the complex dynamics of biomes (The National Academies, 2011). These scientists spent years observing, recording and documenting the flora and fauna. Data collection and analysis has always been the cornerstone of systems research.

Commoditized hardware, telecommunications equipment and free software have enabled production and collection of massive quantities of data. We are surrounded by various sensors (CCTV cameras, RFID tags, mobile phones, GPS devices) that create a flood of data on auto and marine traffic, weather, performance of computing clusters and industrial facilities (NASA, 2011). Internet applications record information about users' actions and allow to voluntarily contribute data. Applications such as Facebook or Twitter have turned their users into human sensor networks that already span the globe (Sakaki et al., 2010; Zhao et al., 2011).

The phenomenon of *increasing data abundance* has become known by a broad term – *"Big Data"* (Waldrop, 2008).

To give the reader a sense of Big Data, consider that a modern gas-powered power plant produces much more data than the New York stock exchange. The International Open Government Dataset Catalog (Tetherless World Constellation, 2011) currently indexes more than 300'000 publicly available datasets covering data on economy, energy, governance, health and public finance. These are two examples of different dimensions of big data: depth and breadth. Governments and agencies around the world have recently started publishing data covering various aspects of their activities. For example, the European Pollutant Release and Transfer Register (E-PRTR) publishes a database on 28'000 industrial facilities in the EU. At the same time the amount of user contributed data, coming from their mobile phones and on-line activities has exploded. For example, the GoodGuide publishes data on more than 115'000 consumer products (GoodGuide, 2011). While E-PRTR data is managed by an agency, GoodGuide's data is contributed by community or "crowd-sourced". Wikipedia (crowd-sourced itself) defines the term as the act of outsourcing tasks, traditionally performed by an employee or contractor, to an undefined, large group of people or community (a "crowd"), through an open call (Wikipedia, 2011b).

4.2 Semantic Web and crowd sourcing

An important aspect of the Big Data movement is the data format. While many datasets are still maintained in tabular fashion (tables and relational databases), increasingly data is published in semantic format. Semantic Web (SW) is a broad term that defines the ecosystem of next generation of internet technologies (Berners-Lee & Hendler, 2001). While there is still debate on which exact implementation and standard will dominate the future of the web (Marshall & Shipman, 2003), there is no doubt that the Semantic Web is likely to prevail due to two enabling aspects: unique resource identifiers and data interoperability. While the current incarnation of the web is about web pages and the links between them, the SW allows for the "Internet of Things" (IoT). The SW allows assigning unique resource identifiers (URI) to all things in the world and defining relations between them; together these consitute the IoT (Gershenfeld et al., 2004). Another important feature of the SW is the interoperability of the data format. Datasets published by different publishers can be combined and reconciled against each other given that all things in the dataset have been uniquely defined (Lassila & Swick, 2011). While within the traditional data formats the records were uniquely identified only within the scope of a dataset, the semantic data format allows for global unique identity. This essentially simple feature provides a mechanism for Big Data to emerge.

The challenge facing the system scientists (also governments and businesses) is to manage the large amount of data and make use of it in decision making. Arguably the way to make sense out of the Big Data is to create tools that would be semantic and allow for collaborative action. Whether collaboration is among a few scientists working on a dataset of wind patterns in Germany or a "crowd" publishing bird sightings around the world, is not relevant. It is important that the datasets are published in the semantic standards and can be shared. Data analysis made possible by massive cloud computing resources and crowds of collaborating scientists and non-academics will help develop more transparent and objective data-driven representations of the world's problems. Employing big data to analyze SoS is a scientific challenge undertaken by the authors. A number of experiments have been performed aiming to integrate big data into decision-support systems, models and simulations. One of the series of experiments deals with the use wikis to enable collaborative big data management and use of that data within simulations[3].

Remembering the stages of SoSe research prescribed by DeLaurentis, wikis can be used in the problem definition phase. Initial analysis can be performed using a wiki environment to to collect, aggregate, curate data, perform queries and let the data structures emerge.

4.3 Wikis for collaborative information management

A wiki is defined as "a website that allows the creation and editing of any number of interlinked web pages via a web browser" (Wikipedia, 2011c). The original creator of the first wiki Ward Cunningham had the goal of creating the simplest on-line database that could possibly work. Wikipedia – the on-line encyclopedia that runs a version of a wiki software – currently hosts pages on 4 million different topics. With that in mind the researchers identify two crucial aspects of wiki-type systems. They are simple to use and they are domain agnostic (generic). In the simplest form, wikis allow multiple users to define concepts and link them together. Besides Wikipedia there are hundreds of wikis covering various subjects, from fictional Pokemon world to the very scientific human genome data. These criteria – simplicity and genericness – have helped the wiki software to establish itself as a novel research support system within information driven research areas, such as biology, pharmacy, genetics and engineering. These criteria make wikis a valuable tool for SoSe research too.

A new approach is offered by the next generation of wiki software, combining the simplicity of the wiki approach with modern semantic technologies. Semantic wikis introduce a powerful feature allowing to mix structured and free-from information (text) within wiki pages. The wiki pages are part web-pages and part database entries. While simple wikis allow one to define concepts and link them together, semantic wikis allow us to define things and relations between them. Systems thinking concerns primarily objects and relations between them and semantic wikis are the perfect tool for managing information about systems. An example of a semantic structure is presented in figure 1.

[3] "Enipedia (`http://enipedia.tudelft.nl`) is an active exploration into the applications of wikis and the semantic web for energy and industry issues. Through this we seek to create a collaborative environment for discussion, while also providing the tools that allow for data from different sources to be connected, queried, and visualized from different perspectives"(Enipedia, 2011)

China

The population of China is [[population::1,331,460,000]].
The capital is [[capital::Beijing]].

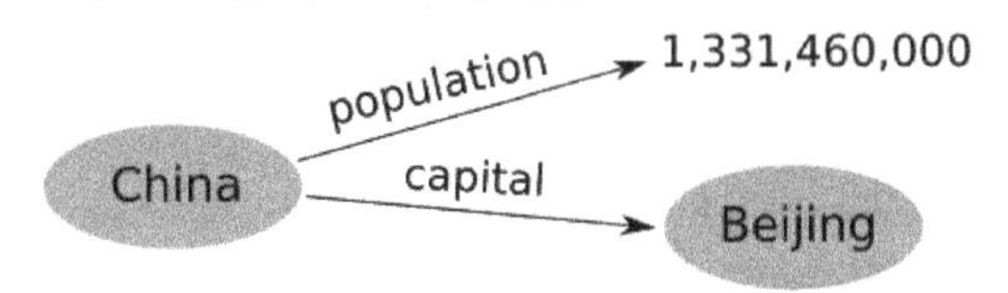

Fig. 1. Semantic wiki – a mix of text and structure

Traditional relational databases have been the dominant decision-support software since the 70s, and have been since then applied in information management across domains. Despite the numerous benefits of relational databases, every database has a unique structure of tables and relations – specific to the domain, environment and problem. Such approach to storing data is called as "structure-first", referring to the fact that tables and relations have to be defined before they can accommodate any data. Moreover, the data structure has to be encoded by professional database developers.

Semantic wiki software offers a different approach, called "data-first". Data can be entered into the system, while simultaneously defining the structure. The fact that the data structure does not have to be fixed beforehand, has tremendous benefits in certain applications. Scientific research, especially interdisciplinary research have embraced the data-first approach. Using data-first approach data structures are part of the continuous research process. In other words the structure of data emerges and evolves as researchers develop better understanding of the problem domain and scope. The evolutionary nature of semantic data structure is especially applicable to managing data in the context of complex, evolving and multi-domain SoSe.

The emergent data structure still allows it to be queried at any point of development. Queries, as in the case of relational databases, are the main method of data analysis. Such functionality allows semantic wikis to assume a new function as domain agnostic (generic) decision-support systems, particularly tailored to interdisciplinary research. The on-line web access, relative ease of use, mix of structured and unstructured data, ability to define and change data structure by the users (not developers) make the wiki approach a valuable tool in ecosystem of the emerging field of big data science.

Another important feature is that any semantic wiki seamlessly ties into the Internet of Things and allows to use and connect to the information entered and maintained by other researchers, agencies, businesses, governments, human volunteers or even machines (Davis et al., 2010). This feature provides semantic wikis with the ability to scale information and knowledge according to the Metcalfe's law (Gilder, 1993), which states that the value of the network is proportional to the number of connected users. Next to this increase of value, Shirky (2008) has talked about the existence of a "cognitive surplus" – the untapped collective mental potential of human society. In other words, it concerns society's spare mental capabilities that go unused in a similar way to computers that are sitting idle. To illustrate his point, he gives a rough calculation that Wikipedia took about 100 million hours of human thought to create. This may seem like an enormous amount of time, but it is roughly the same amount that people in the US spend sitting and watching TV commercials during a single weekend. Finally, Raymond (2001) has observed that "given enough eyeballs, all bugs are shallow". If there is a cognitive surplus, used in a network, we can expect an auto-catalytic increase in quality of networked knowledge and data in semantic wikis.

4.4 Ontologies

One of the challenges often encountered in developing models of complex socio-technical systems stems from the arbitrary boundaries of such systems and the vast number of facts required to conceptualize them. Allenby (2006) argues that the system boundaries are dynamically determined by the query one poses to the system. In other words, the boundaries of the system are determined by the specific research question at hand. For example, studying crime in a city requires different definition of the system than the study of city's water supply.

Using the wikis in the definition phase allows to perform initial analysis of data and create data structures. These data structures can be later translated into system formalisms – ontologies. Ontologies identify things, actors and relationships between them in the system studied. The ontologies already reflect the researchers view of the problem and are part of the abstraction phase withing the SoSe framework.

Mikulecky (2001) emphasizes that complexity manifests itself in the fact that no single system formalization (ontology) can capture all aspects of a complex system. It is a source of much debate how to synchronize or agree on a single ontology of the system. We would argue that such approach is not the most useful. Ontologies are specific to the question at hand and a part of the answer to the research question. Different research questions will require different ontologies but the researcher's tools should allow for flexible mapping of data to ontologies. Semantic wikis allow exactly that.

Ontologies can later be used to create simulations and models of the live system. It is one of the common rules within software development that data structures are more tractable than program logic. Defining simple, clear and transparent ontologies are cornerstone to developing non-trivial agent based models and simulations.

4.5 Agent Based Modeling

Agent Based Modeling (ABM), used in DeLaurentis implementation stage, is a holistic approach, as it provides a perspective on a system from the smallest individual elements to the highest level of system aggregation. While it considers systems in its entirety, it is also reductionist in a sense that it reduces systems to smaller elements if they are fully interrelated with other elements (Bar-Yam, 2003). It is generativist, as it understands systems as a result of continuous process of emergence across multiple levels, starting at the lowest level elements (Epstein, 1999).

In the words of Borshchev & Filippov (2004), the Agent Based approach "is more general and powerful [4] because it enables capturing of more complex structures and dynamics. The other important advantage is that it provides for construction of models in the absence of the knowledge about the global interdependencies: you may know nothing or very little about

[4] than System Dynamics, Dynamic Systems or Discrete Event Simulation

how things affect each other at the aggregate level, or what the global sequence of operations is, etc., but if you have some perception of how the individual participants of the process behave, you can construct the AB model and then obtain the global behavior."

Agent-based models take agents (components) and their interactions as central modeling focus points. Stuart Kauffman has been quoted to say that "an agent is a thing which does things to things" (Shalizi, 2006). Furthermore, Shalizi (2006) states that " An agent is a persistent thing which has some state we find worth representing, and which interacts with other agents, mutually modifying each other's states. The components of an agent-based model are a collection of agents and their states, the rules governing the interactions of the agents and the environment within which they live."

From these interactions, using simple rules and ontologies derived from real data, ABM generate patterns of complex behavior, and serve as a *in silico* experimental device. It should again be noted that ABM are not used to predict the future or identify optima. Their generative nature does allow of to explore *possible futures* through asking what-if type questions.

4.6 Serious games

Karl Jung argued that one of the functions of dreams is to allow the dreamer to practice complex situations and difficult decisions before they happen (Jung & Jaffé, 1989). In that aspect serious games are similar to dreams. They allow the player to practice making decisions in the virtual world. In addition to that they are effective tool for studying, teaching and understanding complex socio-technical systems.

From the systems scientist's perspective serious games are a way to involve humans in simulations. Games have a special power to motivate and instruct (Meadows, 1999). Other advantages are that they can present complex environments, are repeatable, produce high levels of immersion, and are fun (Garris et al., 2002). Serious games provide a basis for organized communication about a complex topic (Duke, 1974; 1980; Kelly et al., 2007), often developed for learning within organizations. Serious gaming has a long history of military purposes and has broadened to a variety of applications, such as business and management science, economics, and inter-cultural communication (Mayer, 2009; Raybourn, 2007). Games are used for education and for the exploration of strategies and policies (Gosen & Washbush, 2004) and, compared to other simulation techniques, games result in a high involvement of the users.

The use of serious games on itself is not sufficient to provide a comprehensive set of insights (Bekebrede, 2010; Bekebrede et al., 2005), therefore, it should not be adopted in isolation. So far, in the literature the combination of serious games and simulation is only adopted as what is now referred to as *simulation games*: serious games with embedded aspects of simulation models. The main disadvantage of games is that there are strong limitations to the complicatedness and length of a game. Even stronger, a conceptually complex game needs to be relatively simple in mechanical terms in order to be effective (Meadows, 1999). Meadows refers to game design, which involves the art and craft of constructing games (Rollings & Adams, 2003). Although there is an elaborate literature on game design for non-educational purposes (cf. Fullerton et al., 2008; Rollings & Morris, 2004; Salen & Zimmerman, 2004; Schell, 2008), there is less literature on serious game design. Several approaches exist, though (cf.

Aldrich, 2004; De Freitas & Oliver, 2006; Frank, 2007; Hall, 2009; Winn, 2009). Essentially, the challenge is to design a game with a good game-*play*, an interesting model of *reality*, and the correct underlying *meaning* (Harteveld, 2011). The opportunity for games to be used together with simulation models is large (Chappin, 2011).

In terms of the DeLaurentis framework, games are another means of implementation, that serves to reconstruct the system analyzed and generate complex phenomena from relatively simple rules.

5. Case studies

Together the Internet of things, semantic information management systems, ABM and games create an ecosystem of data, processes and tools that allow to look at the systems of systems with higher precision and make it easier to find relevant patterns. The primary use of this ecosystem is to advance our understanding of the complex environments and the maturity of our decision making within those environments. The ideas mentioned are not sufficient to change the way we see the world. It also requires commitment from the implied users of these relatively complicated techniques. But recent experience has shown that there are real world interest and applications. Some of this experience is presented further in a form of case studies.

The following case studies demonstrate the use of the tool ecosystem to analyze energy systems.

The first case study concerns the new balancing regime of the natural gas market in the Netherlands and uses the wiki to describe the system and assumptions, later to be used in a simulation. The second case study to analyze the possible future outcomes of the long term development of the European electricity sector, both with an Agent-Based Model and a serious game. Again the wiki [5] is used to gather and define data on thousands of power plants that are later simulated within a complex power market. AgentSpring is the novel ABM simulation framework developed to support the approach already discussed in this chapter and used is in both case studies.

5.0.1 AgentSpring

Before discussing the case studies, it is useful to introduce the modeling framework used in those case studies. Knowing the approach of the framework is helpful in explaining the structure of the models and the terms used.

There are around 60 ABM frameworks in existence, some more popular than others. The motivation for creating another framework was two-fold. Firstly, the framework had to seamlessly integrate the semantic data. Secondly, the framework had to be suited for "super-social" simulations, where behavior of agents is elaborate and diverse. In other words, the framework has to help build models that are data driven and support extensive behavior algorithms.

Surely, these two requirements could be fulfilled by the existing frameworks, provided some modifications were made. But there was also the opportunity to build a framework that would

[5] `http://enipedia.tudelft.nl/`

leverage off the new and powerful open source libraries and changing software development paradigms. AgentSpring gets its name from and makes use of Spring Framework – a popular software development framework, that promotes the use of object oriented software patterns (Johnson et al., 2009). One such pattern calls for separation of data, logic and user interface (Krasner & Pope, 1988). Most modeling frameworks mix the three, which it is a reasonable choice when creating smaller models. However, the separation of concerns (Hursch & Lopes, 1995) and other patterns are especially helpful guidelines for creating models and applications that are sophisticated but transparent.

Another component that AgentSpring brings to modeling is a powerful graph database. A graph database is a database that uses graph structures with nodes, edges, and properties to represent and store information (Eifrem, 2009). The world modeled is created from the ontology that is essentially a graph of objects and their relationships. AgentSpring allows the graph to scale to hundreds of agents, millions of things and relations between them, as represented in figure 2. Such graph databases already power the social networking and other Internet services. The application of graph databases in ABM is new but promising as it allows for more straightforward representation of the system modeled. The graph database makes maintenance of the graph easy and allows to find things and observe patterns by performing pattern matching queries.

Fig. 2. Simulated world: 170'000 agents and things; 650'000 relations between them. Different colors represent different types of relations.

On the conceptual level AgentSpring is inspired by the artificial intelligence classic "Scripts, Plans, Goals, and Understanding: An Inquiry Into Human Knowledge Structures" by Roger C. Schank and Robert P. Abelson. The book suggests that human behavior and understanding of the world is compartmentalized as scripts that are used to execute bigger plans and higher goals (Schank & Abelson, 1977). When executing a plan to go to a restaurant, a person would invoke a script to make reservation in advance, call a cab, perhaps dress up and so on. AgentSpring makes use of the scripts concept to encode agent behavior in a modular way. Agents play their roles in the simulation by executing various scripts. Models are made by combining agents and scripts that define their behavior in the context of social situations. This makes AgentSpring particularly suited to modeling complex socio-technical systems.

AgentSpring decouples agents, their behaviors and their environments making the pieces reusable, composable and easy to manage. Experience has shown that modular and reusable models are the only kind of models that can accommodate changing scope and new research questions.

5.1 Case I: Balancing the natural gas network in the Netherlands

5.1.1 Introduction

As the indigenous natural gas resources near depletion, the goals of the Dutch gas policy have shifted from maximizing state revenues towards energy security and sustainability. It is the intent of government's agenda for the Netherlands to become the gas marketplace of Northwest Europe Ministry of Economic Affairs (2009). In addition to having affordable gas supply the Dutch government hopes to create a liquid gas market and a profitable gas services sector. The new gas balancing regime is another step towards a liberalized gas market in the Netherlands.

In a nutshell, the new balancing regime encourages the market participants to collectively maintain balance of the gas network. The system balance means that the aggregate gas volume entering the system should be equal to the amount of gas leaving the system at any point in time. The system load is determined in the day ahead market, where the market participants submit their gas feed-in and take-off schedules. The individual imbalances are not important as long as the system is in balance as a whole. A power plant operator can consume more gas than scheduled as long as there is someone willing to consume less or supply more at that point in time. If the system goes off balance there are financial penalties introduced to the causers of the imbalance. At the same time the traders contributing to balancing the system are rewarded. Such relatively simple rules could generate interesting aggregate system behavior and complex phenomena might emerge (Bucura et al., 2011).

5.1.2 Model description

An agent based model is constructed to explore the possible effects of the new rules. The operation of the system is simulated to determine the total imbalance, the market participants' cash flows and the natural gas price emerging in the balancing market. Calculating the progression of these variables allows for an ex-ante assessment of the efficiency of the incentive scheme proposed by the new balancing regime and its social cost. Through such modeling and simulation a better understanding of the consequences of the new balancing regime can be obtained.

To arrive at this model the system is decomposed into agents, things and scripts. In the balancing market model initially we distingush between only two types of agents: the system operator overseeing the operation of the gas network, and the gas market participants. Market participants could be traders, gas shippers, power producers, other utility companies. In the context of this initial modeling exercise they are not differentiated. Things represent the physical reality: contracts, capacities, technologies used by the agents. Agents and things are defined in the wiki using the Internet of Things methodology. Every thing or an agent is assigned a unique wiki-page, where the properties and its relations to other things and agents are defined. Figure 3 presents the structure of the model.

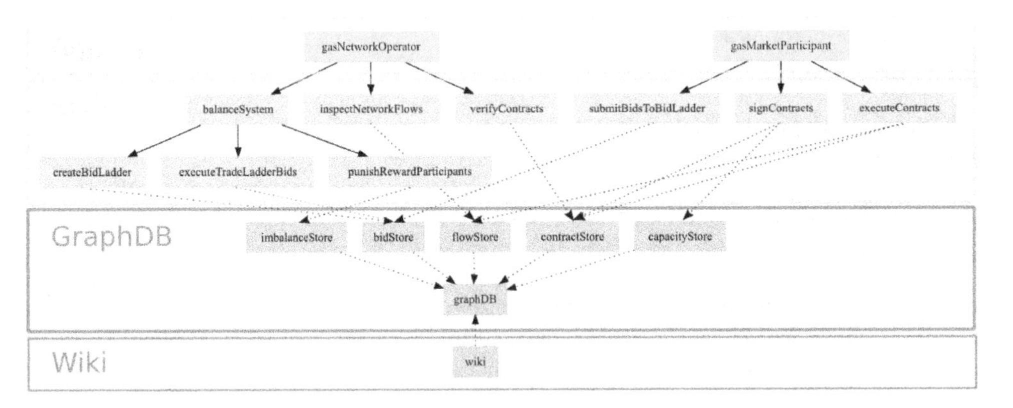

Fig. 3. Model structure

The simulation logic is then decomposed into scripts as discussed previously. The system operator has to make sure the system is in balance during every hour of its operation. The system balancing script is made up three more scripts: *createBidLadder*, *executeBidLadder* and *punishRewardParticipants*. This shows how the scripts can be composed from other scripts and made modular – easier to understand, communicate and maintain. The names of the scripts almost tell the whole story. If the system is out of balance the system operator initiates a secondary market called "bid ladder" (*createBidLadder*), where the imbalance amount is traded in an auction (*executeBidLadder*), balance is restore and the participants are either punished or rewarded (*punishRewardParticipants*). The scope of each script is debatable – one could mold the whole simulation into one big script. But it is most useful when the script contains one piece of simulation logic that is performed by one type of agent and can be well understood and debugged by the modeler.

During the execution of the simulation the agents and things are loaded into a graph database. Scripts are executed within a predefined order: contracts are signed, bid are actioned, gas quantities are delivered – and a complex graph of agents, things and their interactions emerges. The agents acquire knowledge of their environment by querying that graph. They find the cheapest suppliers of gas, the available capacity of the gas transmission network and so forth. The researcher queries the same graph to find patterns in the aggregate system behavior.

5.1.3 Model results

Using the agent based model simulation to explore the operational pathways of the new gas balancing mechanism brought interesting insights. Without going into much detail – which would otherwise require a more dedicated effort and detail explanation of the real system – the results indicate certain pathways of the operation that could lead to increased volatility of the system, higher redistribution of profits and higher social costs of gas network balancing. The research is still in progress and the initial model is being extended to, for example, include heterogeneous behavior of the natural gas traders and different types of contracts.

5.2 Case II: De-carbonization of the power sector

5.2.1 Introduction

Electric power production is largely based on fossil-based combustion, except in environments with abundant hydro-power (IEA, 2008). Fossil fuels have become the lifeblood of developed economies: reducing or replacing their consumption is difficult and expensive. This technology inevitably leads to the emission of carbon dioxide (CO_2), as carbon capture and storage and renewable energy sources are not yet feasible or available on a large scale.Global climate change caused by CO_2 and other greenhouse gases (IPCC, 2007) can be considered a *tragedy of the commons* (Hardin, 1968) for which no effective global coordination, regulation and enforcement has yet been developed. While the cost of abatement is high, doing nothing will eventually be much more expensive (cf. Stern, 2007).

The growing consensus that CO_2 emissions need to be stabilized and then reduced in the course of this century has led to much interest in achieving cost-efficient emission reduction through incentive-based 'carbon policy' instruments – using market signals to influence decision-making and behavior (Egenhofer, 2003) – rather than command-and-control regulation. They need to affect the long term carbon efficiency of the system through investment in new power generation capacity and replacing the old, creating incentives for the "right" investment.

In order to explore the possible effects of the carbon policies we have simulated the complex power generation system of systems (Chappin, 2011; Chappin et al., 2010). The SoS is composed from social systems (power and commodity markets), technical systems (power grids, generation technologies), and the web of relations between the them as illustrated in figure 4.

5.2.2 Model and serious game description

In order to explore the impacts of the policies on the CO_2 emissions of the power generation sector, both a serious game (de Vries & Chappin, 2010)[6] and an agent-based model (Chappin et al., 2010) were developed. In both the model and the game the technical and the social

[6] The serious game is called "Electricity Market Game", and is played online: `http://emg.tudelft.nl`

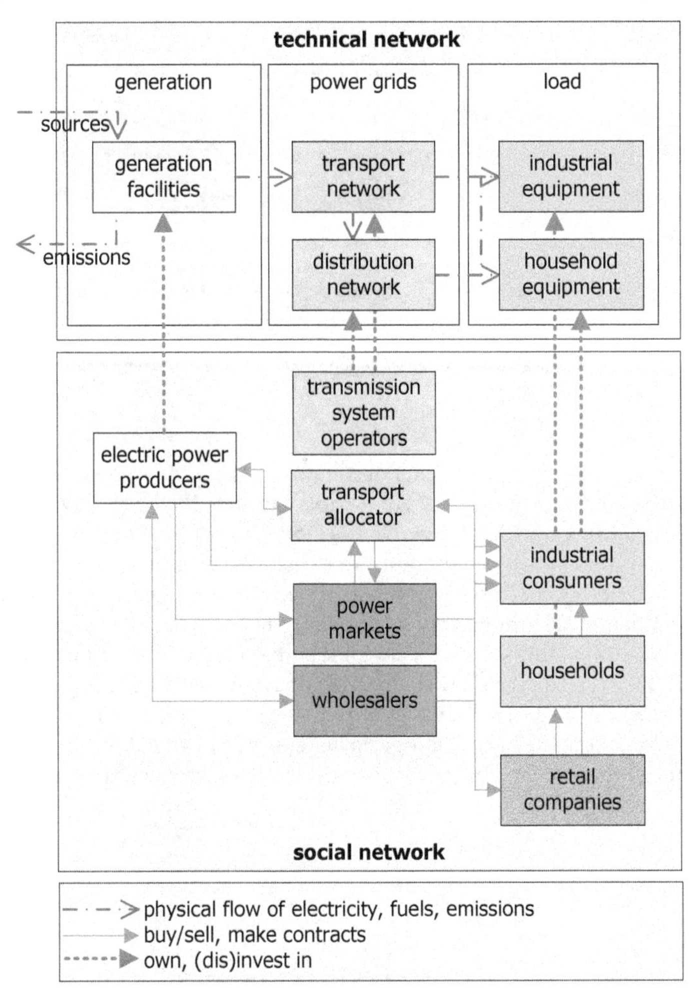

Fig. 4. Socio-technical system-of-systems of electricity production (adapted from (Chappin, 2011))

components of the power system are contained. In the game, people play the roles of the energy producers, which are in the model represented as the agents. Other agents in the model are commodity traders, banks, governments and energy consumers. The game players and the modelled agents interact through markets. A simplified ontology of the system presented in the the model is depicted in figure 5 – a similar ontology is present in the serious game. The model defines multiple commodity markets for coal, natural gas, uranium and biomass, two electricity spot markets, a CO_2 auction, a secondary CO_2 market and a market for power generation technologies. Markets provide a unified mechanism to introduce feedback loops into the model. For example, if many players or agents decide to invest in wind generation technology, the price of the technology in the market may increase, depending on the decision making process of the technology supplier.

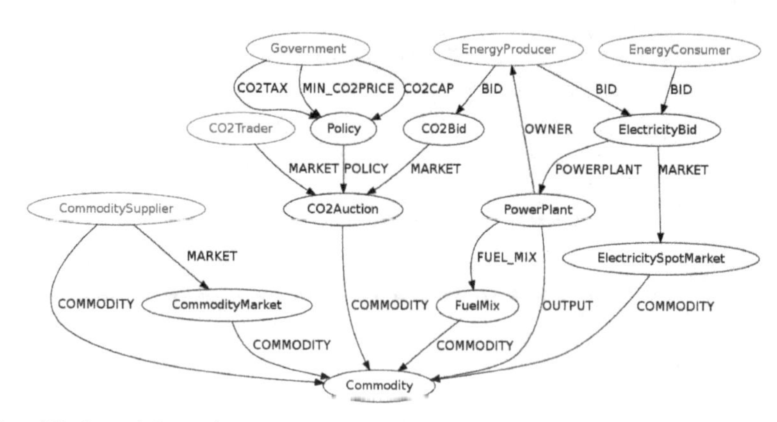

Fig. 5. Simplified model ontology

The decisions and actions of the agents are decomposed using the AgentSpring methodology – into scripts. The diagram in figure 6 lists the scripts executed by the energy producer. Energy producers have the most versatile behavior in the model, they have to operate their generation portfolio, purchase fuels on the commodity markets, sell electricity on the power market, arrange for loans, invest in new power plants and trade on the CO_2 markets. Modular scripts allow to compartmentalize complex agent behavior and allow to develop it piece by piece. Each script concerns only one aspect of agent logic in the context of one function. For example, when trading in the commodity markets the electricity producers are buyers, when trading in the power market they act as sellers, similarly to the commodity traders in the commodity markets. The scripts allow the behavior logic to be reused in different contexts. They allow for more generic and simple algorithms that are easier to understand, maintain and communicate.

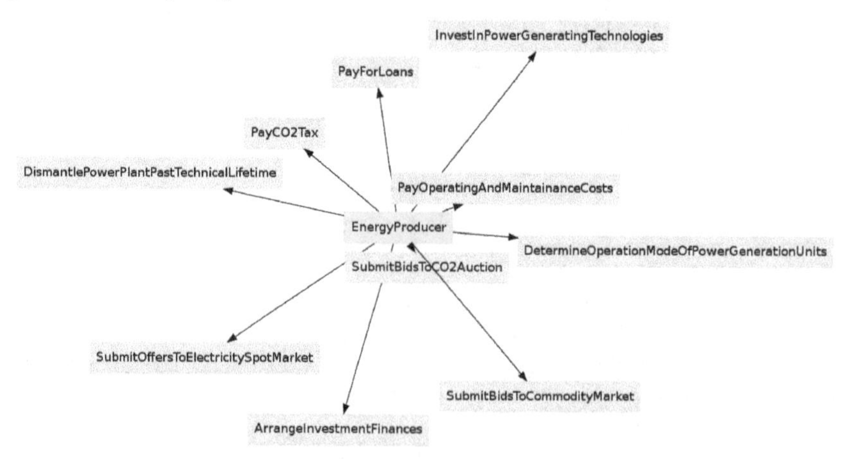

Fig. 6. Energy producer's behavior decomposed into scripts

In the serious game, the players act out their roles, as they have the objective to optimize the value of their company in the long run. Players have to define a strategy to do so and translate their strategy to timely investment and dismantling, and appropriate bidding on markets (de Vries & Chappin, 2010). They are uncertain about future prices and policy instruments.

Crucial for their performance is how players react on such uncertainties and on each other's actions.

This modeling exercise is interesting in the fact that it already uses the Internet of Things to power the model assumptions and scenarios, see figure 7. The power plant data is aggregated from multiple sources (agencies, company data) and includes detailed information on current power generation portfolios of all European countries. The assumptions about technologies, their physical properties are also extracted from EU state of the art specifications. The hour level electricity demand data is also taken from the European Network of Transmission System Operators for Electricity (ENTSOE) databases, converted into a semantic format and used within the model scenario. The wiki in this case is used to aggregate, align and validate the data and re-purpose it for the model.

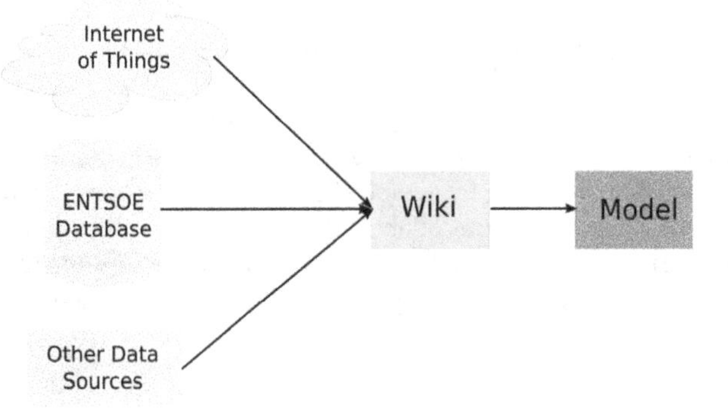

Fig. 7. Using distributed data sources for the simulation

5.2.3 Model results

A key result of the agent-based simulation is that given a certain CO_2 cost – whether through a tax or the price of CO_2 emission rights – carbon taxation leads to lower electricity prices than emissions trading (Chappin et al., 2010). The main reason for this is the difference in investment risk: a tax is more predictable than the market driven CO_2 prices. The uncertainty is factored into the investment decisions via higher discount rates and lead to higher required profitability of the investment. Also the market-based CO_2 prices tend to create investment cycles that induce volatility in the power producer's portfolio. Predictability is a key advantage of taxation, which allows investors to minimize cost over a longer time horizon. In the serious game, similar results have been monitored: if players have the feeling that strict CO_2 policies are in the pipeline they tend to overreact. By playing the game, players tend to be more open to the notion of complex systems and can understand the model faster and deeper (Chappin, 2011). The ecosystem of tools together help in understanding of the evolution – and possibilities of policy design – of the complex system of systems that constitute our electricity infrastructure.

6. Conclusions

The tool and methodology ecosystem described in this chapter are an initial exploration into using big data, collaborative information management to build detailed agent based models of socio-technical systems of systems. It is an iterative process, set in motion by a number of ongoing and planned research projects, with a goal to gain better insights into human behavior and its interaction with the technical and natural environment.

A common misconception about such a research endeavor is that it aims to predict future and will therefore necessarily fail. Instead, the goal of such research is to augment our understanding of reality and make our decision making *less bad*. People will always try to create models of reality and possibly live by them or in them. Heuristics and scripts are useful when we need to make decision with imperfect information. The hope is that with more data, smarter tools and collaborative effort we can reconcile our individual irrationalities into a more objective data-driven understanding of our environment and improve the way we make decisions.

There are number of issues to solve before we are able to enjoy the benefits of this grand vision of data and model augmented decision making. The data is still dispersed in different formats and behind the closed doors of diverse institutions. The different models build by the scientific community are often built in isolation and do not interconnect. Arguably, it is the culture not the technology that is the bottle-neck in progressing systems science. Together with systems engineering we have to do social engineering, in connecting researchers with tools and relevant data, allowing for collaboration and communities to emerge. In order to be effective in analyzing socio-technical systems we have to continuously engineer adequate informational socio-technical systems of systems. This is at the core of the *proactive study of SoSe* proposal.

7. Acknowledgements

This work was supported by the Energy Delta Gas Research program[7], project A1 – Understanding gas sector intra-market and inter-market interactions, by the Knowledge for Climate program[8], project INCAH – Infrastructure Climate Adaptation in Hotspots and by Climate Strategies[9], project Decarbonization of the Power Sector.

8. References

Aldrich, C. (2004). *Simulations and the future of learning: an innovative (and perhaps revolutionary) approach to e-learning.*, Pfeiffer, San Francisco.

Allenby, B. (2006). The ontologies of industrial ecology?, *Progress in Industrial Ecology, an International Journal* 3(1): 28–40.

Bar-Yam, Y. (2003). *Dynamics of Complex Systems*, Westview Press.

Bekebrede, G. (2010). *Experiencing Complexity – A gaming approach for understanding infrastructure systems*, PhD thesis, Delft University of Technology.

Bekebrede, G., Mayer, I., van Houten, S. P., Chin, R. & Verbraeck, A. (2005). How serious are serious games? Some lessons from infra-games, *Proceedings of DiGRA 2005 Conference: Changing Views – Worlds in Play*.

[7] `http://www.edgar-program.com`

[8] `http://knowledgeforclimate.climateresearchnetherlands.nl`

[9] `http://www.climatestrategies.org/`

Berners-Lee, T. & Hendler, J. (2001). Scientific publishing on the semantic web, *Nature* 410: 1023–1024.

Borshchev, A. & Filippov, A. (2004). From System Dynamics and Discrete Event to Practical Agent Based Modeling: Reasons, Techniques, Tools, *Proceedings of the 22nd International Conference of the System Dynamics Society*, pp. 25–29.

Bucura, C., Chmieliauskas, A., Lukszo, Z. & Dijkema, G. (2011). Modelling the new balancing regime of the natural gas market in the netherlands, *Networking, Sensing and Control (ICNSC), 2011 IEEE International Conference on*, IEEE, pp. 68–73.

Chappin, E. J. L. (2011). *Simulating Energy Transitions*, PhD thesis, Delft University of Technology, Delft, the Netherlands. ISBN: 978-90-79787-30-2.
URL: *http://chappin.com/ChappinEJL-PhDthesis.pdf*

Chappin, E. J. L. & Dijkema, G. P. J. (2007). An agent based model of the system of electricity production systems: Exploring the impact of CO_2 emission-trading, *IEEE SoSE: Systems of Systems Engineering*, IEEE, San Antonio, Texas, USA.

Chappin, E. J. L. & Dijkema, G. P. J. (2010). Agent-based modeling of energy infrastructure transitions, *International Journal of Critical Infrastructures* 6(2): 106–130.

Chappin, E. J. L., Dijkema, G. P. J. & Vries, L. J. d. (2010). Carbon policies: Do they deliver in the long run?, *in* P. Sioshansi (ed.), *Carbon Constrained: Future of Electricity*, Global Energy Policy and Economic Series, Elsevier, pp. 31–56. ISBN: 978-1-85617-655-2.

Davis, C., Nikolic, I. & Dijkema, G. (2010). Industrial ecology 2.0, *Journal of Industrial Ecology* .

De Freitas, S. & Oliver, M. (2006). How can exploratory learning with games and simulations within the curriculum be most effectively evaluated?, *Computers & Education* 46(3): 249–264.

de Vries, L. J. & Chappin, E. J. L. (2010). Power play: simulating the interrelations between an electricity market and a CO_2 market in an on-line game, *33st IAEE International Conference, The Future of Energy: Global Challenges, Diverse Solutions*, IAEE, InterContintental Rio Hotel, Rio de Janeiro, Brazil.

DeLaurentis, D. A. & Ayyalasomayajula, S. (2009). Exploring the synergy between industrial ecology and system of systems to understand complexity, *Journal of Industrial Ecology* 13(2): 247–263.

DeLaurentis, D. & Sindiy, O. (2006). Developing sustainable space exploration via system of systems approach, *AIAA Space 2006*.

Duke, R. D. (1974). *Gaming, The Future's Language*, Sage, Beverly Hills, CA.

Duke, R. D. (1980). A paradigm for game design, *Simulation & Games* 11(3): 364–377.

Egenhofer, C. (2003). The compatibility of the kyoto mechanisms with traditional environmental instruments, *in* C. Carraro & C. Egenhofer (eds), *Firms, Governments and Climate Policy: Incentive-Based Policies for Long-Term Climate Change*, Edward Elgar, Cheltenham.

Eifrem, E. (2009). Neo4j - the benefits of graph databases, *no: sql (east)* .

Enipedia (2011). Accessed October 12th.
URL: *http://enipedia.tudelft.nl*

Epstein, J. M. (1999). Agent-based computational models and generative social science, *Complexity* 4(5): 41–60.

Epstein, J. M. & Axtell, R. (1996). *Growing artificial societies: social science from the bottom up*, Complex adaptive systems, Brookings Institution Press; MIT Press, Washington, D.C.

Everett, R. R., Zraket, C. A. & Benington, H. D. (1957). Sage: a data-processing system for air defense, *Papers and discussions presented at the December 9-13, 1957, eastern joint computer conference: Computers with deadlines to meet*, ACM, pp. 148–155.

Forrester, J. W. (1961). *Industrial Dynamics*, Pegasus Communications.

Forrester, J. W. (1968). *Market growth as influenced by capital investment*, Citeseer.

Frank, A. (2007). Balancing three different foci in the design of serious game: engagement, training objective and context, *in* D. Thomas & R. L. Appelman (eds), *Conference Proceedings of DiGRA 2007: Situated play*, University of Tokyo, Tokyo, pp. 567–574.

Fullerton, T., Swain, C. & Hoffman, S. S. (2008). *Game design workshop: a playcentric approach to creating innovative games*, 2nd edn, Morgan Kaufmann, Burlington.

Garris, R., Ahlers, R. & Driskell, J. E. (2002). Games, motivation and learning: A research and practice model, *Simulation & Gaming* 33(4): 441.

Gershenfeld, N., Krikorian, R. & Cohen, D. (2004). The internet of things., *Scientific American* 291(4): 76–81.

Gilder, G. (1993). Telecosm: Metcalfe's law and legacy, *Forbes ASAP* 152: 158–166.

GoodGuide (2011). Accessed September 30.
URL: *http://www.goodguide.com/*

Gosen, J. & Washbush, J. (2004). A review of scholarship on assessing experimental learning effectiveness, *Simulation & Gaming* 35(2): 270–293.

Hall, J. S. B. (2009). Existing and emerging business simulation-game design movements, *Proceedings of ABSEL 2009 annual conference*, ABSEL, Seatle.

Hardin, G. (1968). The tragedy of the commons, *Science* 1968(162): 1243–1248.

Harteveld, C. (2011). *Triadic Game Design – Balancing Reality, Meaning and Play*, Springer-Verlag, London, UK. ISBN: 978-1-84996-156-1.

Hursch, W. L. & Lopes, C. V. (1995). Separation of concerns.

IEA (2008). *World Energy Outlook 2008*, International Energy Agency.

IPCC (2007). *Climate Change 2007: Mitigation of Climate Change Summary for Policymakers*, IPCC, Geneva.

Jamshidi, M. (2005). System-of-systems engineering - a definition.
URL: *http://ieeesmc2005.unm.edu/SoSE_Defn.htm*

Johnson, R., Hoeller, J., Arendsen, A. & Thomas, R. (2009). *Professional Java Development with the Spring Framework*, Wiley-India.

Jung, C. G. & Jaffé, A. (1989). *Memories, dreams, reflections*, Vintage.

Kasmire, J., Chmieliauskas, A. & Boons, F. (Submitted). Changes in meaning: Introducing ideas and tools for exploring changes in word meanings, *Journal of Machine Learning Research* .

Kelly, H., Howell, D., Glinert, E., Holding, L., Swain, C., Burrowbridge, A. & Roper, M. (2007). How to build serious games, *Communications of the ACM* 50(7): 45–49.

Krasner, G. E. & Pope, S. T. (1988). A cookbook for using the model-view controller user interface paradigm in smalltalk-80, *J. Object Oriented Program.* 1(3): 26–49.

Lassila, O. & Swick, R. R. (2011). Resource description framework (RDF) model and syntax. Accessed September 30.
URL: *http://www.w3.org/TR/WD-rdf-syntax*

Los Angeles Times (2011). Accessed September 30.
URL: *http://latimesblogs.latimes.com/technology/2011/03/the-apple-ipad-2-has-been-selling -out-in-stores-nationwide-since-it-launch-last-friday-and-the-scarcity-of-the-device-is-li.html*

Marshall, C. C. & Shipman, F. M. (2003). Which semantic web?, *Proceedings of the fourteenth ACM conference on Hypertext and hypermedia*, ACM, pp. 57–66.

Mayer, I. S. (2009). The gaming of policy and the politics of gaming: A review, *Simulation & Gaming* 40(825): 825–862.

Meadows, D. H. (2008). *Thinking in Systems: A Primer*, Chelsea Green Publishing. ISBN: 978-1603580557.

Meadows, D. L. (1999). Learning to be simple: My odyssey with games, *Simulation & Gaming* 30(3): 342–351.

Mikulecky, D. C. (2001). The emergence of complexity: science coming of age or science growing old?, *Computers and Chemistry* 25(4): 341–348.

Ministry of Economic Affairs (2009). Government Report – The Netherlands as a Northwest European Gas Hub.
URL: *http://www.apxendex.com/uploads/Corporate_Files/APX_Quarterly/Government_Report_-_The_Netherlands_as_a_Northwest_European_Gas_Hub.pdf*

MIT (1953). The sage air defense system.
URL: *http://www.ll.mit.edu/about/History/SAGEairdefensesystem.html*

NASA (2011). Planetary skin project. Accessed September 30.
URL: *http://www.planetaryskin.org/*

Nikolic, I. (2009). *Co-Evolutionary Method For Modelling Large Scale Socio-Technical Systems Evolution*, PhD thesis, Delft University of Technology. ISBN 978-90-79787-07-4.

Pei, R. S. (2000). Systems of systems integration (sosi)-a smart way of acquiring army c4i2ws systems, *Proceedings of the Summer Computer Simulation Conference*, pp. 574–579.

Raybourn, E. M. (2007). Applying simulation experience design methods to creating serious game-based adaptive training systems, *Interacting with Computers* 19: 206–214.

Raymond, E. S. (2001). *The cathedral and the bazaar: Musings on Linux and open source by an accidental revolutionary*, O'Reilly & Associates, Inc. Sebastopol, CA, USA.

Rollings, A. & Adams, E. (2003). *Andrew Rollings and Ernest Adams on Game Design*, New Riders Games. ISBN 978-1592730018.

Rollings, A. & Morris, D. (2004). *Game architecture and design: a new edition*, New Riders, Indianapolis.

Sakaki, T., Okazaki, M. & Matsuo, Y. (2010). Earthquake shakes twitter users: real-time event detection by social sensors, *Proceedings of the 19th international conference on World wide web*, ACM, pp. 851–860.

Salen, K. & Zimmerman, E. (2004). *Rules of play: game design fandamentals*, The MIT Press, Cambridge.

Schank, R. C. & Abelson, R. P. (1977). *Scripts, plans, goals and understanding: An inquiry into human knowledge structures*, Vol. 2, Lawrence Erlbaum Associates Hillsdale, NJ.

Schell, J. (2008). *The art of game design: a book of lenses*, Morgan Kaufmann, Burlington.

Shalizi, C. R. (2006). Methods and techniques of complex systems science: An overview, *arXiv.org* arXiv.org:nlin/0307015.
URL: *http://www.citebase.org/abstract?id=oai:arXiv.org:nlin/0307015*

Shirky, C. (2008). Here comes everybody, *Web 2.0 Expo*, San Francisco, USA.
URL: *http://www.shirky.com/herecomeseverybody/2008/04/looking-for-the-mouse.html*

Stern, N. (2007). *The Economics of Climate Change: The Stern Review*, Cambridge University Press. ISBN 05-21-70080-9.

Tetherless World Constellation (2011). Accessed September 30.
URL: *http://logd.tw.rpi.edu/demo/international_dataset_catalog_search*

The National Academies (2011). Accessed September 30.
URL: *http://www7.nationalacademies.org/archives/International_Biological_Program.html*

Waldrop, M. (2008). Big data: wikiomics, *Nature* 455(7209): 22–25.

Wikipedia (2011a). Accessed September 30.
URL: *http://en.wikipedia.org/wiki/Moscow_âĂŞ_Saint_Petersburg_Railway*

Wikipedia (2011b). Accessed September 30.
URL: *http://en.wikipedia.org/wiki/Crowdsourcing*

Wikipedia (2011c). Accessed September 30.
URL: *http://en.wikipedia.org/wiki/Wiki*

Winn, B. M. (2009). The design, play, and experience framework, *in* R. E. Ferdig (ed.), *Handbook of research on effective electronic gaming in education*, Vol. III, Information Science Reference, Hershey, pp. 1010–1024.

Zhao, S., Zhong, L., Wickramasuriya, J. & Vasudevan, V. (2011). Human as real-time sensors of social and physical events: A case study of twitter and sports games, *Arxiv preprint arXiv:1106.4300* .

6

Future Intelligent Earth Observing Satellite System (FIEOS): Advanced System of Systems

Guoqing Zhou
[1]*Guilin University of Technology, Guilin,*
[2]*Department of Modeling, Simulation and Visualization Engineering,*
Old Dominion University, Norfolk, VA,
[1]*China*
[2]*USA*

1. Introduction

Diverse natural disasters such as earthquakes, volcanoes, tornadoes, subsidences, avalanches, landslides, floods, wildfires, volcanic eruptions, extreme weather, coastal disasters, sea ice and space weather, tsunami, pollution events, have ravened many lives and damaged a number of properties each year in our home planet, resulting in imposing heavy burden on society [13]. For example, a deadly 8.0 Ms Wenchuan earthquake occurred at 14:28:01.42 on May 12, 2008 in Sichuan province of China has killed at least 69,197, injured 374,176, made 18,222 missed, and gave rise to about 4.8 million people homeless; a 9.0 magnitude earthquake occurred on the seafloor near Aceh in northern Indonesia on 26 December 2004, 00:58:53 UTC, causing a huge tsunami wave, hitting the coasts of Indonesia, Malaysia, Thailand, Myanmar, India, Sri Lanka, Maldives and even Somalia in Africa, resulting in over 280,000 people lost their lives. The town of Lhoknga, near the capital of Banda Aceh, was completely destroyed by the tsunami.

It has been demonstrated that the losses of life and property from natural disaster can be reduced through analysis of earth observing data acquired by spaceborne. However, not all of disasters, such as tsunami, earthquake can so far be warned and predicted in advance, consequently, scientists have spent enormous efforts to exhaustively seek for thread of these complex natural phenomena from earth observing system in order to develop predictive measures so that people have enough time to prepare, plan, and response these disasters. Unfortunately, little progress has been made due to the lack of adequate measurements and the depth with which we fully understand the physics of these phenomena [9], i.e., the current measurements and observations largely cannot meet the demands of the disaster warning.

In order to increase our ability to monitor and predict natural disasters, Zhou *et al.* [18] in the early 2000 presented an envisioned architecture, named *"future intelligent earth observing satellite system (FIEOS)"*. Afterwards, Bayal *et al.* [5] and Habib *et al.* [10] presented the similar concepts. The FIEOS would substantially increase intelligent technologies into Earth observing system in order to improve the temporal, spectral, and spatial coverage of the area(s) under investigation and knowledge for providing valued-added information/data products to users. The envisioned FIEOS is especially significant for people, who want to learn about the dynamics of, for example, the spread of forest fires, regional to large-scale air quality issues, the spread of the harmful invasive species, or the atmospheric transport of volcanic plumes and ash [9]. The FIEOS is honored as advanced global earth observing system of systems (GEOSS). This paper attempts to state *(i)* what the challenges of "traditional" GEOSS are; *(ii)* how FIEOS (advanced GEOSS) increases the efficiency of monitoring natural disaster, to improve the natural disaster management, and to mitigate disasters; *(iii)* how the FIEOS can enhance our understanding to Earth system and Earth progress; *(iv)* how the FIEOS significantly benefits the disaster reduction through, for example, real-time response to time-critical natural disaster.

2. "Traditional" Global Earth Observing System of Systems (GEOSS) and Challenges

In order to improve our capability of understanding of the Earth system, and enhance prediction of the natural disaster, an agreement for a 10-year implementation plan for a Global Earth Observation System of Systems, known as *GEOSS,* was reached by member countries of the Group on Earth Observations at the Third Observation Summit held in Brussels on February 16, 2005. The GEOSS was envisioned as an international cooperative effort to bring together existing and new technologies in hardware and software, making it all compatible in order to share data and information worldwide at no cost. All subscribing nations maintain their independent role in developing and maintaining the system, collecting data, analyzing data, enhancing data distribution, etc. (www.codata.org/GEOSS/capetown-meeting.html). With such an envisioned architecture, the GEOSS is anticipated to meet the need for timely, quality long-term global information for sound decision making for, and enhance delivery of benefits to society relating to disaster preparedness.

Meanwhile, the US 10-year Strategic Plan for the U.S. Integrated Earth Observation System was publicly presented by the President's Science Advisor on September 8, 2004. The Plan addressing nine societal benefit areas include [19]:

- Improvement of weather forecasting
- Reduction of loss of life and property from disasters
- Protection and monitor our ocean resources
- Understanding, assessment, prediction, mitigation and adaption to climate variability and change
- Support of sustainable agriculture and combination of land degradation
- Understanding of the effect of environmental factors on human health and well-being
- Development of the capacity to make ecological forecasts
- Protection of and monitoring of water resources
- Monitoring and management of energy resources

The characteristics of the above strategies lie in as follows.

1. Improve current capability, needs, and deficiencies of satellite/air/ground imaging systems in order to integrate global systems, joint data collection, and behavioral modeling initiatives, etc.,
2. Improve data collection to increase our understanding of how disasters evolve and to assess current phenomena of natural disaster,

3. Improve data processing capability and techniques and data/information visualization techniques in order to develop better models to make prediction model more accurate,
4. Establish global disaster reduction & warning system in order to make the results more precise, and
5. Establish national all-disasters emergency communication system in order to make all disaster information accessible and warn citizens with consistent, accessible, and actionable messages.

The GEOSS is promising. However, because of the complexity of Earth processes and mechanism of natural disaster, the requests for earth observing system has shifted from previous imaging mode, spatial resolution, spectral resolution, revisit capability, etc. to on-board data processing, event-driven data collection, value-added products, etc. (see Figure 1). This means that a much advanced earth observing system is needed.

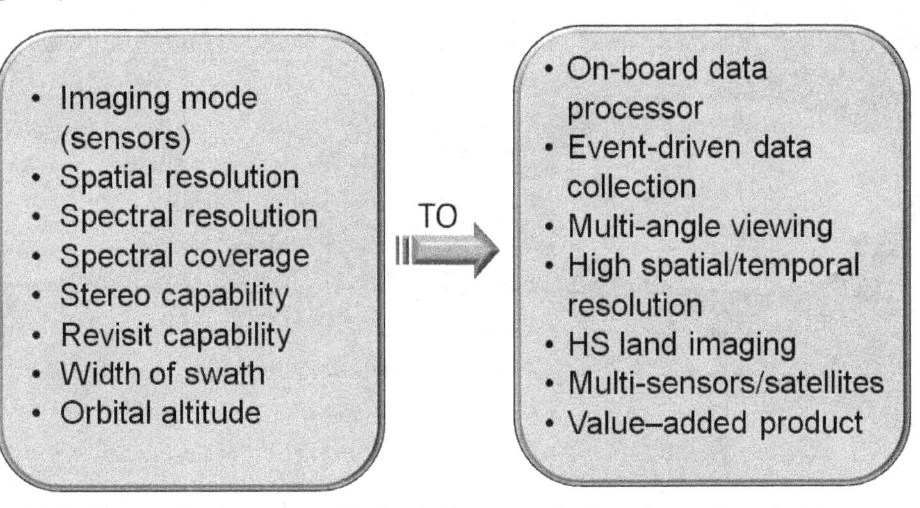

Fig. 1. The features for disaster monitoring in future earth observing system of systems

3. Future Intelligent Earth Observing Systems - FIEOS

3.1 Architecture of intelligent Earth observing satellite system - FIEOS

As mentioned above, the GEOSS has its limits in rapid response to time-critical natural disaster. Thereby, a much advanced earth observing system, called future intelligent earth observing system was proposed by Zhou in 2000 [18]. The envisioned FIEOS is a space-based architecture for the dynamic and comprehensive on-board integration of Earth observing sensors, data processors and communication systems. The implementation strategies suggest a seamless integration of diverse components into a smart, adaptable and robust Earth observation satellite system to enable simultaneous, global measurements and timely analyses of the Earth's environment for a variety of users (Fig. 2). The architecture consists of multiple layer networked satellites. Each EO satellite is equipped with a different sensor for collection of different data and an on-board data processor that enables it to act autonomously, reacting to significant measurement events on and above the Earth. They collaboratively work together to conduct the range of functions currently performed by a few large satellites today through the use of high performance processing architectures and reconfigurable computing environments [1], [3-4]. The FIEOS will act autonomously in controlling instruments and spacecraft, while also responding to the commands of the user interested to measure specific events or features. So, users can select instrument parameters on demand and control on-board algorithms to preprocess the data for information extraction. All of the satellites are networked together into an organic measurement system with high speed optical and radio frequency links. User requests are routed to specific instruments maximizing the transfer of data to archive facilities on the ground and on the satellite. Such an earth observing system allows measurement from *in situ*, air borne or space based sensors to be multiple practical usage that can help in making critical decisions for societal benefits.

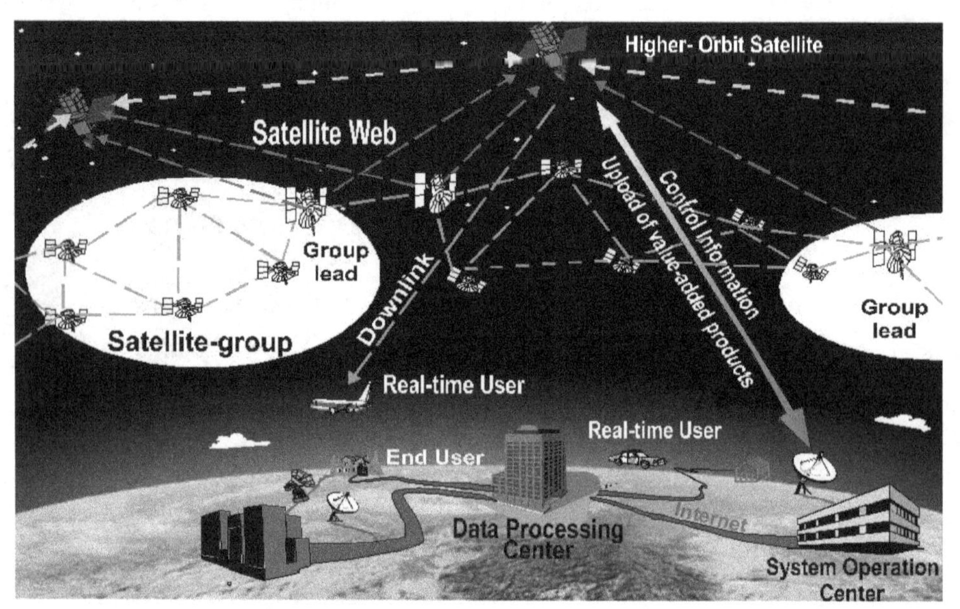

Fig. 2. The architecture of a future intelligent earth observing satellite system (courtesy of Zhou et al. [18])

3.2. Event-driven Earth observation

The optimum earth observing system to meet the specific needs and mandates on specific and achievable societal benefits never stops [10]. A called *event-driven observation* in FIEOS has been presented [18]. The operational mode is that each EO sensing system independently collects, analyzes and interprets data using its own sensors and on-board processors. When a sensing system detects an event, e.g., a forest fire, the sensing-satellite rotates its sensing system into position and alters its coverage area via adjusting its system parameters in order to bring the event into focus [13]. Meanwhile, the sensing-satellite informs member-satellites, and the member-satellites adjust their sensors to acquire the event, resulting in a multi-angle, -sensor, -resolution and -spectral observation and analysis of the event (Fig. 3). These data sets are merged to a geostationary satellite according to the changes detected. Meanwhile, the geostationary further processes the data to develop other products, e.g., predictions of fire extend after 5 days, weather influence on a fire, pollution caused by a fire, etc. These value-added products are then transmitted to users. The details of the event-driven Earth observation can be referenced to Zhou et al. [18].

Fig. 3. Event-driven earth observation

4. Disaster Reduction from FIEOS

The significant characteristic of FIEOS is its capability to rapid response to time-critical disaster, relative to the GEOSS. Thus, the FIEOS largely benefits to both decisions makers and the general public for disaster prediction and disaster preparedness. As an example, this paper describes how the FIEOS improve weather forecast for reduction of disaster.

4.1 Reduction of disaster through improvement of weather forecast

The improvement of weather forecast will largely enhance prediction of disasters caused by extreme weather, such as flooding, landslide, etc. Although weather satellite observing system, along with the other associated national and international data management mechanisms, is probably most mature relative to other observing systems, the improvement of accuracy of weather-forecasting, the enhancement of observations (e.g., wind and humidity profiles, precipitation), the improvement of long-term weather forecasting, and the access and delivery of essential weather forecast products to user for meeting requirements of timely short- and medium-term forecasts are still urgently essential for natural disaster reduction [6].

The shortcoming of the current earth observing system is that its spatial, temporal-, and spectral resolution and sensing capability cannot obtain sufficiently high accurate, gridded worldwide weather [6], resulting in that the different weather users, such as real-time, mobile users, cannot dynamically access the desired data in an near instantaneous and global access manner. The envisioned FIEOS observing system is capable of providing users to near instantaneously access to worldwide weather data for a given point, a path, or an area in time and space anywhere in the world via satellite broadcast or direct send/receive satellite link. Especially, FIEOS provides the weather forecasting data with different levels of scales: macro-meso-micro level. At the macro-scale level, users, such as commercial airlines pilot, can obtain weather forecasting information from forecast centers via wireless. At the meso-level, user can directly obtain weather forecasting products from a forecast or data processing center via either wireless or wire access. Alternatively, the user can also gain access to the database(s) described weather information to generate his or her own weather products using wireless/wire user software. For those mobile users, including truck drivers, farmers, and private car owners, they can receive the broadcast weather information directly from the forecasting information center using hand-held device. The devices can also be designed to have a direct send/receive satellite transmission capability, and the broadcast center may be local TV, universities, and radio stations, etc.

4.2 Dissemination to lay user

The obvious shortcoming of the current earth observing system is that the lay users cannot actively be involved. Relatively, one of the benefits of FIEOS lies in its broad range of user communities, including managers and policy makers in the targeted societal benefit areas, scientific researchers, engineers, governmental and non-governmental organizations and international bodies. In particular, FIEOS would serve lay users who directly receive satellite data (in fact, the concept of data means image-based information, rather than traditional remotely sensed data) using their own receiving equipment. The operation appears to the end-users as simple and easy as selecting a TV channel by using a remote control (Fig. 4). Moreover, the authorized users are allowed to upload the user's command for accessing and retrieving data via on-board data distributor according to the user's requirement and position [18]. In this fashion, a lay user on the street is able to use a portable wireless device to downlink/access the satellite information of his surroundings from satellite or from the Internet. Homes in the future are also able to obtain atmospheric data from the satellite network for monitoring their own environments. The FIEOS will enable people not only to see their environment, but also to "shape" their physical surroundings.

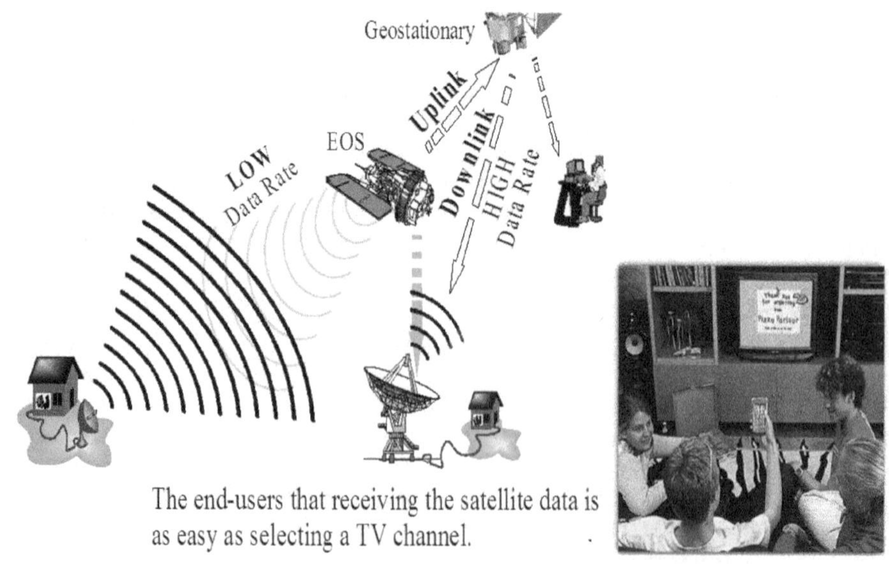

Fig. 4. Lay-user receive the satellite inforamtion just like selecting a TV channel.

5. Conclusions

Although significant advances in our ability to measure and understand the Earth system and process using earth observing system have been obtained, the emergency task should is to build an intelligent, comprehensive, integrated, and sustained earth observation system, despite remaining multiple technical challenges, to ultimately realize a wide range of disaster reduction. This paper presented an envisioned FIEOS which is intended to enable simultaneous, global measurements and timely analyses of Earth's environments for a variety of users through dynamic and comprehensive on-board integration of Earth observing sensors, data processors and communication systems.

FIEOS provides the nation with a unique and innovative perspective on the intelligent observing system for disaster reduction for (1) reducing losses of life and property; (2) improving weather forecasting; and (3) disseminating information to lay users. Realization of FIEOS, which is a much advanced GEOSS, is an exciting opportunity to make lasting improvements in delivering prediction of disaster to our people.

6. Acknowledgement

This paper is financially supported by China Natural Science Foundation (CNSF) under the contractor number of 41162011.

7. References

[1] Alkalai, L, 2001. An Overview of Flight Computer Technologies for Future NASA Space Exploration Missions, 3rd IAA Symposium on Small Satellites for Earth Observation, April 2 - 6, Berlin, Germany.

[2] Allen, T., G. Lu, and D. Wong. (2003). Integrating remote sensing, terrain analysis, and geostatistics for mosquito surveillance and control, ASPRS Annual meeting, Alaska, May 5-9, 2003.

[3] Armbruster, P. and W. Wijmans, 2000. Reconfigurable on-board payload data processing system developments at the European Space Agency, ESA-presentation at SPIE 2000, Volume 4132-12.

[4] Bergmann, N.W. and A.S. Dawood, 2000. Reconfigurable Computers in Space: Problems, Solutions and Future Directions, The 2nd Annual Military and Aerospace Applications of Programmable Logic Devices (MAPLD'99) Conference, Laurel, Maryland, USA.

[5] Baysal, Oktay, and Guoqing Zhou, 2004. Layer Users can share satellite image information in future? XXth ISPRS Congress, Geo-Imagery Bridging Continents, Special Session-FIEOS, Istanbul, TURKEY, 12-23 July 2004, DVD.

[6] Cicerone, Ralph, et al., 2001. Climate Change Science: An Analysis of Some Key Questions, National Research Council, 2001, National Academy Press, pp 23-24.

[7] U.S. Air Force, 1994. SPACECAST 2020 Technical Report, Volume I, Prepared by the Students and Faculty of Air University, Air Education and Training Command, United States Air Force, Maxwell Air Force Base, Alabama, June 1994

[8] Dutton, John, 2002. Opportunities and Priorities in a New Era for Weather and Climate Services, Bulletin of the American Meteorological Society, September 2002, vol. 83, no. 9, pp. 1303-1311.

[9] Habib Shahid and S. J. Talabac, 2004. Space-based Sensor Web for Earth Science Applications - An integrated Architecture for providing societal benefits, XXth ISPRS Congress, Geo-Imagery Bridging Continents, vol. XXXV, part B1, 12-23 July , Istanbul, Turkey, (DVD)

[10] Habib, Shahid and P. Hildebrand, 2002. Sensor Web Architectural Concepts and Implementation Challenges - An Heuristic Approach, SPIE International Symposium on Remote Sensing, Crete, Greece, September 22-27, 2002.

[11] Krishna Rao, Susan J. Holmes, Ralph K. Anderson, Jay S. Winston, and Paul E. Lehr, Weather Satellites: Systems, Data, and Environmental Applications (American Meteorological Society, Boston, 1990), 7-16.

[12] Schoeberl, M., J. Bristow and C. Raymond, 2001. Intelligent Distributed Spacecraft Infrastructure, Earth Science Enterprise Technology Planning Workshop, 23-24 January.

[13] Strategic Plan, 2004: Strategic Plan for the U.S. Integrated Earth Observation System, (released in September 2004) at U.S. http: //iwgeo.ssc.nasa.gov/draftstrategicplan/ieos_draft_strategic_plan.pdf

[14] NASA, 20004. Technical Activity Reports for the Nine Societal Benefits areas at http://iwgeo.ssc.nasa.gov/documents.asp?s=review

[15] United Nations (U.N.) Population Division, 1996. World Population Prospects 1950-2050, The 1996 Revision, on diskette (U.N.,New York,1996).

[16] Van der Vink, et al., 1998. Why the United States is Becoming More Vulnerable to Natural Disasters, EOS, Transactions, American Geophysical Union, vol. 79, no. 44, November 3, 1998, pp. 533-537.

[17] Weiher, Rodney [ed], 1997. Improving El Niño Forecasting: The Potential Economic Benefits, NOAA, U.S. Department of Commerce, 1997, p.29, p.43, for U.S. agriculture and fisheries, respectively.

[18] Zhou, G., O. Baysal, and J. Kaye, 2004. Concept design of future intelligent earth observing satellites, International Journal of Remote Sensing, vol. 25, no. 14, July 2004, pp. 2667-2685.

[19] NSTC Committee on Environment and Natural Resources, Interagency Working Group on Earth Observations, Strategic Plan for The U.S. Integrated Earth Observation System, www.whitehouse.gov/sites/default/files/microsites/.../eocstrategic_plan.pdf

Permissions

The contributors of this book come from diverse backgrounds, making this book a truly international effort. This book will bring forth new frontiers with its revolutionizing research information and detailed analysis of the nascent developments around the world.

We would like to thank Dr. Adrian V. Gheorghe, for lending his expertise to make the book truly unique. He has played a crucial role in the development of this book. Without his invaluable contribution this book wouldn't have been possible. He has made vital efforts to compile up to date information on the varied aspects of this subject to make this book a valuable addition to the collection of many professionals and students.

This book was conceptualized with the vision of imparting up-to-date information and advanced data in this field. To ensure the same, a matchless editorial board was set up. Every individual on the board went through rigorous rounds of assessment to prove their worth. After which they invested a large part of their time researching and compiling the most relevant data for our readers. Conferences and sessions were held from time to time between the editorial board and the contributing authors to present the data in the most comprehensible form. The editorial team has worked tirelessly to provide valuable and valid information to help people across the globe.

Every chapter published in this book has been scrutinized by our experts. Their significance has been extensively debated. The topics covered herein carry significant findings which will fuel the growth of the discipline. They may even be implemented as practical applications or may be referred to as a beginning point for another development.

The editorial board has been involved in producing this book since its inception. They have spent rigorous hours researching and exploring the diverse topics which have resulted in the successful publishing of this book. They have passed on their knowledge of decades through this book. To expedite this challenging task, the publisher supported the team at every step. A small team of assistant editors was also appointed to further simplify the editing procedure and attain best results for the readers.

Our editorial team has been hand-picked from every corner of the world. Their multi-ethnicity adds dynamic inputs to the discussions which result in innovative outcomes. These outcomes are then further discussed with the researchers and contributors who give their valuable feedback and opinion regarding the same. The feedback is then collaborated with the researches and they are edited in a comprehensive manner to aid the understanding of the subject.

Apart from the editorial board, the designing team has also invested a significant amount of their time in understanding the subject and creating the most relevant covers. They scrutinized every image to scout for the most suitable representation of the subject and create an appropriate cover for the book.

The publishing team has been involved in this book since its early stages. They were actively engaged in every process, be it collecting the data, connecting with the contributors or procuring relevant information. The team has been an ardent support to the editorial, designing and production team. Their endless efforts to recruit the best for this project, has resulted in the accomplishment of this book. They are a veteran in the field of academics and their pool of knowledge is as vast as their experience in printing. Their expertise and guidance has proved useful at every step. Their uncompromising quality standards have made this book an exceptional effort. Their encouragement from time to time has been an inspiration for everyone.

The publisher and the editorial board hope that this book will prove to be a valuable piece of knowledge for researchers, students, practitioners and scholars across the globe.

List of Contributors

G. Reza Djavanshir and Ali Alavizadeh
Johns Hopkins University, USA

M.J. Tarokh
K.N. Toosi University of Technology, Iran

Takafumi Nakamura
Fujitsu Fsas Inc., Japan

Kyoichi Kijima
Tokyo Institute of Technology, Japan

Ben Clegg and Richard Orme
Aston Business School, United Kingdom

Angelo Corsaro
PrismTech, USA

Douglas C. Schmidt
Vanderbilt University, USA

Alfredas Chmieliauskas, Emile J. L. Chappin, Chris B. Davis, Igor Nikolic and Gerard P. J. Dijkema
Delft University of Technology, The Netherlands

Guoqing Zhou
Guilin University of Technology, Guilin, China
Department of Modeling, Simulation and Visualization Engineering, USA
Old Dominion University, Norfolk, VA, USA

Printed in the USA
CPSIA information can be obtained
at www.ICGtesting.com
JSHW011325221024
72173JS00003B/66